Printing Technologies
and Applications

A multidisciplinary roadmap

Online at: https://doi.org/10.1088/978-0-7503-2568-4

Printing Technologies and Applications

A multidisciplinary roadmap

Susanne Klein
Brimstree Studio, Stroud, UK

Carinna Parraman
Centre for Fine Print Research, University of the West of England, Bristol, UK

IOP Publishing, Bristol, UK

ISBN   978-0-7503-2568-4 (ebook)
ISBN   978-0-7503-2566-0 (print)
ISBN   978-0-7503-2569-1 (myPrint)
ISBN   978-0-7503-2567-7 (mobi)

DOI   10.1088/978-0-7503-2568-4

Version: 20241001

IOP ebooks

British Library Cataloguing-in-Publication Data: A catalogue record for this book is available from the British Library.

Published by IOP Publishing, wholly owned by The Institute of Physics, London

IOP Publishing, No.2 The Distillery, Glassfields, Avon Street, Bristol, BS2 0GR, UK

US Office: IOP Publishing, Inc., 190 North Independence Mall West, Suite 601, Philadelphia, PA 19106, USA

To my parents, Ursula and Heinrich Klein, who did not want me to study physics, but art. The book is a compromise.

Contents

Acknowledgements

I would like to thank all my PhD students who taught me physics and chemistry and my print making colleagues who were always patient with my approach to printing.

Author biographies

Susanne Klein

Susanne Klein is a fellow of the Institute of Physics and a printmaker. In 1995 she became a Royal Society Research Fellow at the University of Bristol where she worked on 19th century optics. From 1998 until 2017 she was employed at Hewlett Packard Labs and specialised in liquid crystal display technology and new materials for 3D printing. From 2018 until 2024 she was an EPSRC Manufacturing Fellow at the Centre for Print Research with the task to resurrect forgotten photomechanical printing methods of the 19th century. Her main interests are colour capture, colour reproduction and continous tone photomechanical printing processes.

Her artistic practice focuses on the reproduction of memories and dreams with an emphasis on colour.

She exhibits regularly nationally and internationally and has been curating the printmakers' exhibition at the University of the West of England for the last six years. She speaks regularly at international conferences about her scientific research and her artistic practice.

She is a committee member of the IoP's special interest group Printing and Graphics Science, a committee member of the Imaging Science group of the Royal Photographic Society and a member of the British Liquid Crystal Society's awards committee.

Carinna Parraman

Dr Carinna Parraman is Professor of Design, Colour and Print and Director of the Centre for Print Research at the University of the West of England, Bristol. At the CFPR, she leads a multi-disciplinary research team of academics and technicians with a range of expertise: from scientists and technologists to craftspeople, designers and artists and has just completed a large four-year project funded by Research England (E3).

Carinna is an artist and printmaker, and her practice spans the art and science of colour printing, halftoning, colour perception and appearance. She collaborates with a range of external stakeholders including industry, business and technology, cultural heritage, galleries and museums.

Printing Technologies and Applications
A multidisciplinary roadmap
Susanne Klein and Carinna Parraman

Chapter 1

Fundamentals of printing

1.1 What is printing?

The word 'print' triggers the idea of image or text. 'A book is in print' means that it is available and for sale, even though many books are no longer produced as physical objects of ink on paper. Posters decorate student rooms and framed prints are common wall decorations in first homes and holiday lets. Gifts are wrapped in printed paper and street signs would not exist without print. Print does not come to mind immediately when we look at floors, mobile telephones, cars, doors, walls, computers, food, and drink, but it is everywhere—everywhere where we need many of the same and of the same quality.

Print began with money (figure 1.1). Minting a coin was the earliest form of 3D or 2.5D printing. To make a coin an image in combination with text is debossed again and again into metal. The imprint turns it into currency, making it recognizable and sometimes more valuable than the metal itself. Without being one of many, a coin does not have a defined value. Print is not only the method of production, but also defines a coin's function and is a guarantee of consistent quality.

However, print still has the reputation of being inferior to hand-made, one-off, images, whether on canvas, paper, crockery, or textiles. Why? Producing a print requires technical skills and investment from the crafts-person, artist, or technologist beyond what is needed for a one-off.

Let's look at a pencil drawing in comparison to a relief print, something achievable at home.

1. **Upfront investment**

    Drawing: a piece of paper and a pencil or pen are needed.

    Relief print: a pencil, a piece of wood or linoleum, at least one knife, paper, ink, a roller, and a printing press (which can be as simple as a spoon) are needed.

    It is already clear that the investment into the production of a print is much higher than that into a drawing.

doi:10.1088/978-0-7503-2568-4ch1

Figure 1.1. Alexander the Great, silver coin, 336–323 BC.

2. **Skills**

   Drawing: the only skill needed to produce a mark on paper is that the creator knows how to use a pencil.

   Print: the creator needs to know how to make a mark on the substrate using a pencil, cut the relief into the substrate following the drawing, ink the plate, transfer the ink to the paper via pressing the paper onto the plate, and finally pull the print.

3. **Time investment**

   Drawing: As little as a few seconds.

   Print: At least four times the amount of time needed for a drawing.

Even though 'monoprint', when only one print can be and is pulled from a plate, is practised as an art form, it becomes clear that the training, financial, and time investment into print demands that more than one print is pulled. Beyond the minimum number of prints which justifies the investment, print becomes financially viable.

The challenge for the printmaker or printing process is to produce results of consistent quality, which is for example a problem for 3D printing.

We have not included 3D printing in this book, because it is better defined as rapid prototyping—it does not work using a plate, die, or mould, it needs post-processing, it is mostly not printed on a substrate which stays part of the final print, and it struggles to produce the same quality for each print.

We have included non-impact printing. Non-impact printing, such as inkjet for example, can be seen as being related to 3D printing as it does not work with a printing plate, but it is printed on a substrate and does not need any post-processing.

Printing can replace repetitive processes and enhance quality. Let us consider two examples: crockery and circuit boards.

Hand-painted crockery is valued highly. The aim of the production process is to generate the same pattern on each piece of a table service. To achieve this aim the craftswoman (it is mostly done by women) paints the pattern on only one part of the set, for example only on diner plates or only on cups. The colour choice, painting technique, and motif are all decided by the designer. The craftswoman must repeat the same brush strokes again and again. The slightest deviation leads to a devaluation of the object. The work is monotonous, requires the executing person to concentrate for long periods, and can lead to repetitive stress injuries. The process is easily replaced by a decal process where the image is printed on a tissue substrate and then transferred onto a ceramic object. The transfer can be done by hand or machine. Not only can the same quality of image reproduction be achieved more easily, but more complicated patterns become possible.

Figure 1.2 shows an antique china Blue Willow meat serving plate, one of the first designs made using a decal from an engraved copper plate.

Circuit boards (figure 1.3) are essential in our daily life. They are in almost all electronic devices and appliances. The traditional production of circuit boards is a multi-step process, of about 18 steps, including several printing and etching steps. The application of inkjet printing has replaced five steps in the production of etched circuit boards, making the production more cost efficient [1]. Inkjetting of conducting inks on flexible substrates has led to the invention of new products: from electronic skin patches using printed batteries to printed photodetectors for warehouse management. Sensors printed on stretchable substrates increase user comfort in healthcare applications.

Figure 1.2. 1869 Hulse & Adderley Blue Willow chinoiserie transferware stone china serving platter.

Figure 1.3. A circuitboard in a laptop.

Security printing is another visible, but often overlooked, sophisticated application. It involves the printing of marks on packaging, banknotes, certificates, stamps, art objects, pharmaceuticals etc, which identify the object as genuine and prevent forgery, tampering, or counterfeiting. A security print increases the value of a product and installs confidence in its quality.

A modern banknote (https://www.bankofengland.co.uk/banknotes/polymer-10-pound-note), more sophisticated than the example shown in figure 1.4, contains a multitude of printed security features which make counterfeiting time-consuming and expensive. It is printed on a polymer substrate and has the following security features [2]:

- A printed hologram which changes between the number 10 and pound symbol when tilted.
- The print on the see-through window is gold on the front and silver on the back.
- A portrait of Queen Elizabeth II is printed on the see-through window with '£10 Bank of England' twice around the edge.
- A printed colour changing quill at the side of the see-through window.
- A 3D image of the coronation crown on the front of the note above the see-through window.
- In the same place on the back of the note is a book shaped copper foil patch with the letters 'JA'.
- The print of 'Bank of England' and '10' is raised on the front of the note.
- Microprint underneath the Queen's portrait showing the value of the note.
- Under UV light the number 10 will appear in red and green on the front of the note.
- A unique printed serial number.

Figure 1.4. Reichsbanknote from 1922, overprinted with a new value due to inflation. (Reproduced with permission from Nicolai von Poncet.)

Digital technology allows one to record the features of the banknote to a certain extent, but it cannot be printed from this information without major investment into different printing technologies. We have come full circle. After a period when every household had a non-impact printer and prints became a commodity with a lifespan of only hours, printmaking has returned to the hands of specialists. Today hardly anything, maybe the odd boarding pass, is printed at home. Photographic images are mostly consumed digitally, but the customer is ready to spend £100 or more on a photomechanical print for a special occasion, from christenings to Christmas. Most of the clothes we wear have at least one printed component—the washing and brand label. Hundreds of companies offer customized garments. T-shirts and face masks are printed on demand based on customers' design. Without print we would not know what is in the tin when we buy it, whether we can still eat the food which has spent half a lifetime in the fridge, or where we are supposed sit in the cinema.

The lifetime of a print can be seconds—a receipt from a supermarket—or hundreds of years, like a woodcut by Albrecht Dürer (figure 1.5). It all depends on how the ink attaches to the substrate and that is what this book is about.

1.2 What is a print?

Printing (or imprinting) is closely tied to the need to communicate and share knowledge, and in Europe was developed through the efforts of pioneer printers spread across several countries. The history of modern printing begins in the fifteenth century, but who invented printing at that time was highly contested and controversial. Attempts of the century was succinctly summarized, 'Holland has books but no documents, France has documents but no books, Italy has neither books nor documents, while Germany has both books and documents.' [3]

In 1440 Johannes Gutenberg of Mainz pioneered the casting of moveable metal type and, by adapting a screw thread printing press, revolutionized the reproduction, dissemination, and mass communication of the printed word. The print-workshop of

Figure 1.5. Albrecht Dürer, *Die vier apokalyptischen Reiter* (*The Four Horsemen*), 1498. (This image has been obtained by the author from the Wikimedia website where it was made available by Brwz (2019) under a CC0 licence. It is included within this chapter on that basis. It is attributed to Brwz.)

Mainz produced the first Latin Bible, which was originally called the Mazarine Bible, then the 42-line Bible, and is now generally known as the Gutenberg Bible (figure 1.6). The first book to be printed in England was by William Caxton in 1477.

For many centuries the primary method of relief printing remained the same, which involved inking and printing from a raised surface of wooden type, or wood engraved or woodcut images. This method evolved to using metal plates that were either etched or engraved.

Stepping back further, an early example of relief printing was the Chinese printing method using stone, which originated as a means of copying and disseminating inscriptions on stone. Rubbings were made from the raised areas and the carved or indented lines remained white.

The printed image in the sixteenth century radically shifted the balance of knowledge from hearsay and description to a way of distributing visual information and illustration of what things looked like. Francis Bacon (1561–1626) highlighted that knowledge needed to be practical and relevant for mankind. Printed media enabled a knowledge and social revolution for a large under-privileged sector, spreading images to the mass market, and presenting new insights and learning that had at the time been accessible to only the very few.

From the sixteenth century onwards printing steadily expanded to include artworks as well as books and documents. In the mid-sixteenth century, printers

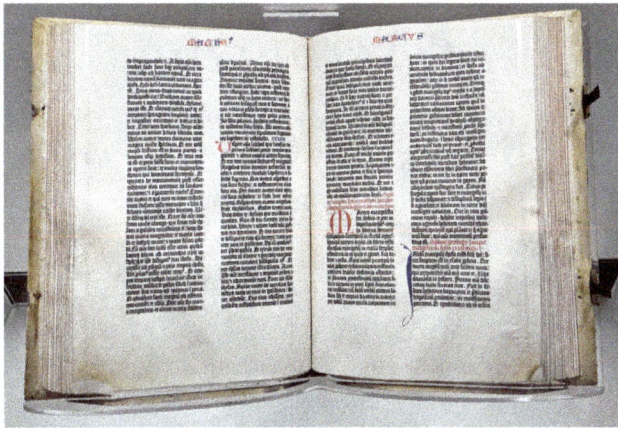

Figure 1.6. A Gutenberg Bible on display at the US Library of Congress. This Gutenberg Bible remains on public display in the Great Hall of the Jefferson Building. (This image has been obtained by the author from the Wikimedia website where it was made available by MC BSU (2024) under a CC0 licence. It is included within this chapter on that basis. It is attributed to MC BSU.)

also began to explore metal-engraving and letterpress. Etching and engravings of paintings and drawings were highly sought after and prized in the print market. Woodcuts and wood-engravings flourished, including the experimentation of multi-block chiaroscuro prints. A large number of woodblock prints made in the mid-sixteenth century were designed solely for the print market.

Printing in the seventeenth century generally concentrated on the spread of information through handbills, pamphlets, newspapers, and acts of parliament. As different typefaces in Europe began to increase and be shared, there were also attempts to standardize the design of type. Joseph Moxton, an English type-founder and printer, provided detailed instructions for printers on the construction of type founding, composition, and hand-press printing.

Across the eighteenth century, during a period of intense scientific revolution, the gathering, storing and retrieving of knowledge, as demonstrated in printed ency-clopaedias, was crucial to furthering the knowledge of the philosopher, the artist, the industrialist, and the scientist. These included Joseph Moxon's *Mechanick Exercises: Or The Doctrine Of Handy-Works* (originally printed as a series in 1683 and 1685 and then published as a book in 1703), the two-volume *Cyclopaedia; or A Universal Dictionary of Arts and Sciences* by Ephram Chambers (1680–1740), was published in England in 1728, and Denis Diderot's 28 volume *Encyclopédie, ou Dictionnaire raisonné des sciences, des arts et des metiers* [4] published in 1772 (figure 1.7). The *Encyclopédie* fuelled a culture of intellectual debate in the so-called Enlightenment salons in France that brought together discursive activity from different social and cultural backgrounds.

It is interesting to note that in the early nineteenth century the mechanics of printing had progressed so little that a fifteenth century printer working in the print-workshop of Mainz would still be familiar with the technology.

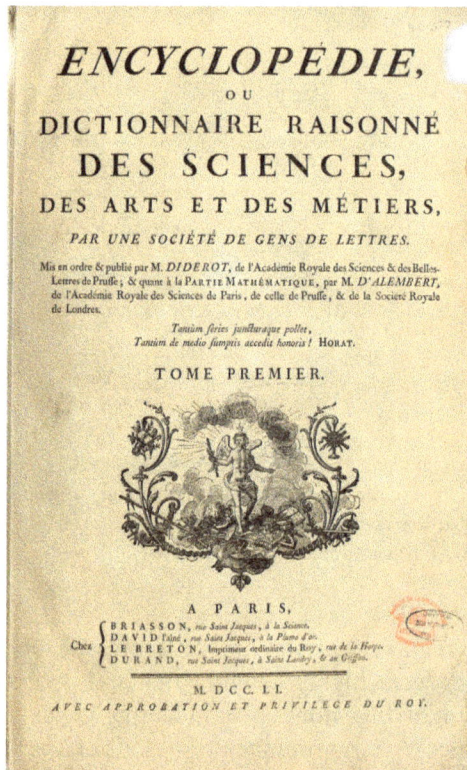

Figure 1.7. Cover of the *Encyclopédie, ou dictionnaire raisonné des sciences, des arts et des métiers* (*Encyclopaedia or a Systematic Dictionary of the Sciences, Arts and Crafts*) edited by Denis Diderot and Jean le Rond D'Alember. (This image has been obtained by the author from the Wikimedia website where it was made available by a Long List of Contributors to the Encyclopédie and Jaybear (2013). It is stated to be in the public domain. It is included within this chapter on that basis. It is attributed to Jaybear.)

The next significant development of the nineteenth century was the production and refinement of paper, as invented by the Foudrinier brothers in 1803, Koenig's steam powered printing machine, and photography and photomechanical printing as invented by Daguerre and Fox Talbot.

The first significant developments of note were the improvements to the flatbed printing press by Charles Stanhope in 1800, who replaced wood with a press made entirely of iron with the addition of a series of iron levers. The iron significantly increased the rigidity and evenness of the print, and the levers enabled a greater amount of pressure, thus saving human exertion. The Stanhope press was quickly used by publishers and newspaper printers including *The Times*.

Other presses that were built to similar specifications included the Columbian press (1817) and the Albion press (c. 1822) ([5], p 51).

Paper, until this time, was made by hand as single sheets in woven wire mesh moulds. The Fourdrinier machine replaced the individual moulds by a continuous feed on a mesh belt, and paper was pressed, smoothed and dried using large, heated cylinders. The width of the rollers became the standard size for paper. Although the continuous web of paper was introduced in 1803, continuous printing did not arrive until some 40 years later when rotary printing was introduced by William Nicholson Hoe in 1843.

The first steam powered printing machine was invented by Frederick Koenig (1774–1833). Koenig went on to make many improvements and developments, including a powered cylinder press with automatic inking, and a machine that could print on both sides of the sheet. The main factor that limited the speed of printing was less to do with the machine itself, but rather the need to hand-feed paper by the sheet. William Nicholson Hoe's six-cylinder rotary printing press meant that sheets could be printed much faster than on a flat bed. Known as the Hoe web perfecting press, it used a continuous roll of paper and printed on both sides of a page in a single operation.

From artistic to commercial exploitation and mechanization, photographic illustration through the agency of light represented a bridge between reproduction by hand, to reproduction by photochemical and then photomechanical means [6]. In the eighteenth century, drawing apparatus enabled artists to create profile portraits by illuminating their sitter by a spotlight, and with the paper placed behind them a silhouette was achieved by outlining the shadow. Examples included the Prosopographus or the Automation Artist, invented by Charles Samuel Hervé (c. 1785–1866) around 1820, in which the likeness of the sitter could be captured by tracing with a pencil. Other drawing machines, such as the Pantograph, were used as a replicating or enlarging device, and also as a tool for engraving into metal.

The English potter and manufacturer Josiah Wedgewood (1730–1795) and chemist Humphrey Davy experimented with light and a light-sensitive solution of silver nitrate. They explored contact printing by placing objects on a light-sensitive paper and silhouette portraits by painting a light-sensitive background and illuminating the sitter against the background. However, they were not able to make the images permanent. It was William Herschel in 1819 who discovered the best method to fix images using sodium thiosulphate (incorrectly named at the time as hyposulphite of soda) (figure 1.8).

The nineteenth century also marked a huge separation and demarcation in the trades. Illustrations made in wood were considered as low art due to their ubiquity, until Thomas Bewick reinvigorated the craft.

In the late eighteenth century Paul Sandby, a watercolourist, etched and engraved his own plates, and Thomas Bewick made his own wood-engravings. William Morris and William Hooper at Kelmscot press believed in the integrity of the craft, and William Blake's own plates evolved from his own practice and experimentation. They were all involved into the making of plates, blocks and printing. In the eighteenth century all illustrations had to be engraved in copper, which was costly to produce, and had to be printed separately from the text. Wood engraving became the main process for illustration in ordinary books, thus leading to a range of quality and their ubiquity throughout the nineteenth century (figure 1.9). From the 1820s lithography and steel engraving began to be used for illustration in books.

Figure 1.8. John Herschel's first glass-plate photograph, dated 9 September 1839, showing the mount of his father's 40 foot telescope. (This image has been obtained by the author from the Wikimedia website where it was made available by Svajcr (2009). It is stated to be in the public domain. It is included within this chapter on that basis. It is attributed to Svajcr.)

Mid-nineteenth century scientists and photographic inventors were experimenting with chemicals and light to fix images to paper. These new photographic methods fundamentally changed the way that the reflection of light and fine details of a scene could be captured. The process was based on the blackening of silver salts on a negative as a result of sunlight on the activated areas, which are converted to black metallic silver. The blackened areas then inhibited light from passing through to the plate or paper beneath. However, the problem faced by early photographers was the ability to fix an image and to prevent it from fading. Realizing the fleeting nature of photography, printers started to experiment with plates and presses instead of sensitized paper.

One of the most significant achievements of the nineteenth century was the ability to reproduce photographs by means of a printing press. Joseph Nicéphore Niepce made the first permanent photograph around 1822, and the first photogravure plate in 1826. He coated a pewter plate with a light-sensitive bitumen. The plate was exposed to a transparent paper negative containing an image. The plate was cleaned, and areas of the plate protected by the dark areas of the image were etched in acid to create an intaglio plate, which could then be printed. He exposed his plate under a line engraving and once etched he was able to print it as a photographic illustration.

The bitumen process was further refined by Fox Talbot, who discovered that a combination of gelatine and bichromate of potash, after exposure to light, became

Figure 1.9. Woodcut press, from an engraving in *Early Typography* by William Skeen, Colombo, Ceylon, 1872. (This image is stated to be in the public domain.)

insoluble and impermeable to an etching solution. By coating a copper or steel plate and exposing it under a negative, it became possible to etch a plate and thus created a photographic illustration. Talbot's method formed the basis of modern photogravure (figure 1.10), as developed by Karl Klic in 1897.

To transfer an image into a reproducible image on paper required a mechanical halftone, which can also be described as a series of evenly spaced dots but varying in size from small to large, that when viewed at a distance gave the illusion of a continuous tone image. The first experiments into mechanical halftone photogravure were made by Fox Talbot in 1853, where he placed a piece of gauze, or as he described it a photographic veil, between a paper negative and a plate. In 1858 he patented this as the photoglyphic method. Fox Talbot's demonstration of the use of light-sensitive bichromated gelatine and platemaking was the beginning of a wealth of successful methods for continuous tone photomechanical printing with remarkable tonal range and image quality.

Figure 1.10. Photogravure of Victor Hugo by Comte Stanislaw Julian Ostrorog dit Walery (1830–1890). (This image has been obtained by the author from the Wikimedia website where it was made available by the French Government, Ministry of Culture and Mschlindwein (2004). It is stated to be in the public domain. It is included within this chapter on that basis. It is attributed to Mschlindwein.)

Printers found that by combining print and photography they could not only produce an image that did not fade, but also an image could be reproduced many times and at speed. Considered as an 'unacknowledged revolution' [7], this fundamental shift was brought about by etchers and engravers, originally trained as goldsmiths, who brought about a 'standardized knowledge' [8], where a repeatable image was effectively created by transferring ink from a printing plate made of wood or metal onto a sheet of paper using a variety of techniques.

Fast forward to the twenty-first century, print production is certainly fundamental to our everyday lives, and yet it is still considered as one of the more utilitarian of the consumer markets that have had an impact on all aspects of our lives. The UK printing industry is the fifth largest in the world and contributes almost £775 million to the trade balance with a turnover of £13.8 billion, employing c. 116 000 people. Customer demand for bespoke and innovative printed materials is driving the industry to extend its market and explore high-value printing, for example, coloured patterns and print-on-demand for fashion and design, safer and legible packaging for food products, organic printed materials and bio-sensors in medicine, and complex and hidden features for banknotes and security cards. Each of the techniques requires specialist methods of printing. Therefore, how a print is created—the type of print technology, the ink and substrate—has a particular impact on the design process and the outcome.

Technologies appropriate for printing text and numbers are less useful/applicable for printing images. Printing that includes both colour, photographic images, and type is a relatively recent development of the digital age. Think back to the 1980s and desktop publishing, where all three aspects could first be combined and printed seamlessly.

Born out of a drive for improvement, history has demonstrated that technical innovation has evolved alongside image and design production. Many of the early nineteenth century processes have become redundant as more efficient solutions were sought and found. In some cases, where high-quality images ware required, speed and expediency superseded quality and longevity. But why this renewed interest in printing now, at the beginning of what can be considered as a screen-based, digitally mediated century? As humans we still need to find practical solutions to physical problems, and while many issues can be addressed through AI and computer modelling, we are still in need of the craft skills from the past. For example, of interest is William Henry Fox Talbot's role in developing a method for printing photographic images using ink, paper and a printing press, which resulted in a printed image that did not fade. Many photomechanical variations quickly evolved from Talbot's original method, and within 25 years the print industry was able to adapt this knowledge commercially for high-speed printing of items such as cardboard boxes, and high-value printing for items such as banknotes and postage stamps. However, what is less known is how Fox Talbot made those breakthroughs. The aim of this book is to explore the overlooked, historical and alternative aspects of print.

In the twenty-first century we have many opportunities to access images. Using an online word search of possibly the world's most famous painting, the *Mona Lisa*, reveals hundreds upon thousands of images. Some may be instantly recognizable as the original *Mona Lisa*, other images may have been cropped, modified, stylized in different ways, or have been appropriated for other uses such as printing on t-shirts, bags and umbrellas. I have used the term 'original' in this context, based on my assumption that most of us will be familiar with Leonardo da Vinci's *Portrait of La Gioconda*, which sits behind bullet-proof glass in the Louvre, Paris (http://www.louvre.fr/en/explore/the-palace/from-the-mona-lisa-to-the-wedding-feast-at-cana).

Technology in the twenty-first century has the capacity to reproduce all art, whereby audiences of millions can access images simultaneously on many different platforms, all of which has a significant impact on the public's viewing, under-standing, and consumption of images.

Although the impact of digital technology has brought the gallery to our homes, a more profound impact on the way artworks could be viewed, accessed and disseminated occurred 100 years ago. This occurred through the combination of

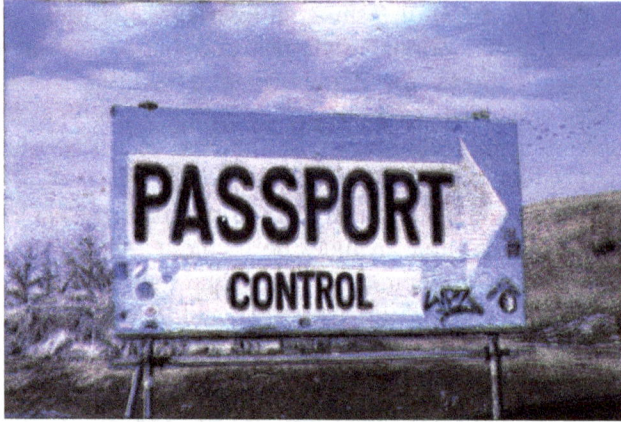

Figure 1.11. Full colour Woodburytype, the first commercially successful photomechanical printing process in the nineteenth century. (Print by Susanne Klein.)

photographic processes and the printing press, whereby photographs of objects, artworks and scenes could be made and committed faithfully to paper using inks, plates and photomechanical reproductive methods (figure 1.11).

The Internet has provided access to unlimited images and things we may never have anticipated or known about. Digital technologies have irrevocably changed and challenged the way we look at, construct, and print images and objects. We work digitally and incorporate numerous digitally aided technologies as a part of our daily workflow. As we move from real-world texture to screen-based or printed representations of texture, our understanding and engagement is mediated by the screen or printed matrix. Furthermore, colour-printing technologies have evolved from coloured dots on paper to coloured dots on three-dimensional objects. What has remained unchanged are the thin-film CMYK process inks and colours using which images are printed. The next step in colour printing is a modification to the thickness of the film to create a new dimensionality or functionality.

1.3 Different printing technologies

Many books, articles and papers have been written on printing processes and their innumerable sub-methods. To assist the non-expert reader, we have presented this section as an introductory review of the different printing processes, including a short history, definitions, and applications in industry today.

In the bibliography, we have shared our key texts, spanning both history and technology, which can direct the reader to more detailed literature [6, 9–14].

1.3.1 Relief printing

1.3.1.1 History
Relief may be the oldest printing method. An early example of relief printing was the Chinese method of printing from stone, which originated as a means of copying and disseminating inscriptions on stone. Rubbings were made from the raised areas, and

the carved or indented lines remained white. The crusades in the eleventh, twelfth and thirteenth centuries opened Europe to oriental culture and technology. By then, woodblock print on textiles was well established in India and China. In the early 1300s it appeared in Europe, followed by woodcut printing for playing cards and religious iconography [15].

Most people will associate relief printing with the advent of book printing in Europe and Gutenberg's Bible in the 1450s. The art of book printing evolved into an industrial, high-volume activity for five centuries and was the primary method of producing printed text.

Now commonly known as letterpress, in the twentieth century relief printing was overtaken by offset lithography in the printing of books and newspapers. In the twenty-first century it is considered an endangered craft, however, letterpress printing has experienced a revival as an art form.

Today relief printing is used for packaging and labels [16] and letterpress has been replaced by flexography.

1.3.1.2 Definition

The principle of relief printing is that the image is carried on the highest parts of the plate, see figure 1.12. The surface is uniformly raised from the plate, and the ink is applied to the raised areas. Areas which should not be printed are cut away. In German, it is called *Hochdruck*, which means 'high print'.

1.3.1.3 Materials

The printing plate can be made from different materials. The most common ones are wood and metal, but stone, clay, linoleum and even dough are used. The following techniques belong to relief printing and are named after the material of the printing plate:

- Woodcut (the wood for the plate is cut parallel to the grain. The structure of the wood will be visible in the final print).
- Wood engraving (the wood for the plate is cut perpendicular to the grain. The structure of the wood will not be visible in the final print).
- Stone engraving.

Figure 1.12. Relief print: the ink sits on the highest part of the plate.

- Linocut.
- Metal cut (the material of the plate can be copper, steel, or lead).
- Letterpress.
- Relief etching (invented by William Blake c. 1788). The white background is etched away. For example, magnesium plates are used for letterpress.
- Rubber stamping.

1.3.1.4 Process
Woodcut, wood engraving, stone engraving, linocut, and relief etching are still practised by artists. The plates are cut mainly by hand. Sharp cutting tools and profiled chisels are used to incise and remove unwanted areas from the block. A modern tool is the laser cutter which can cut plates or rollers. To laser cut an image, an intermediate digital step is necessary. The image must be prepared as a vector file. As in hand-engraved plates, the non-ink carrying areas are cut away. Figure 1.13 is an example of the process. A photographic image was rendered into an image suitable for relief by applying the filter 'mezzotint' in Photoshop. The image was then copied into Illustrator, from which it was sent to a Trotec laser cutter. After cutting, the plate was cleaned, inked, and inserted into an Albion letterpress. Paper was put on top of the plate and the image was transferred by pressure.

1.3.1.5 Contemporary methods of relief printing and their applications
The raised image is generated on flexo plates either by a photographic process on a photopolymer plate, by direct laser etching of a metal plate, or through a moulding process of an elastic material. For example, a light-sensitive photopolymer plate is exposed through a halftone image on transparent film. The unexposed/unhardened sections of the plate are washed away, leaving the exposed/hardened sections as a relief image on the plate, which is then mounted on a printing cylinder. The mounted

Figure 1.13. A laser cut relief plate (right) and the print pulled from it (left).

plates are inked, and the ink is transferred by pressure onto the final substrate. The printing speed can be as high as 750 m min^{-1}.

Its main application is labelling and packaging, especially for food cartons and wrappers [17]. A wide variety of inks can be used, and the choice of ink is governed by the application and the intended substrate, for example, whether the packaging must be suitable for food.

1.3.2 Intaglio printing

1.3.2.1 History

Intaglio printing, which began as engraving, is almost as old as relief printing. It was developed in the fifteenth century based on the refined art of engraving practised by gold and silversmiths. It is believed that designs were recorded by inking the engraved objects and imprinting on paper [13]. It was only a small step to use engraved plates as printing plates. Albrecht Dürer (1471–1528), who learned from metal engravers and translated their technique into print, was an early master of the technique.

The process also evolved at the same time as Gutenberg was printing his Bible and in the exact same location, Mainz, an unidentified 'Master of the Playing Cards' cut the earliest known copper plates. It could even be that he belonged to Gutenberg's workshop [15].

The intaglio technique has always been intended for pictorial applications and was perceived as vastly superior to the sometimes crude relief plate. It forms the largest of the family of processes which are described below.

1.3.2.2 Definition

Intaglio is a printing process that describes a range of cutting or etching techniques by incising a plate with a sharp instrument or acid to create an image. This time the image is generated by the lower parts of the printing plate see figure 1.14. In German, it is called *Tiefdruck*, which means 'low print'.

1.3.2.3 Process

The etching process employs a sharp needle onto a copper or zinc plate, which is coated with an acid-resistant wax ground. The needle is used to draw lines through the ground, and create outlines, detail or cross-hatching to create tones. As the plate is dipped into acid, the acid bites into areas of the surface that are scratched by the

Figure 1.14. Intaglio print: the ink sits in the depressions of the plate and is pulled out by capillary forces during printing.

needle. The wax ground is removed from the plate. During the inking process, ink is pushed into the groves, and surplus ink is wiped away. Paper is pressed onto the surface and runs through an etching press under pressure. The ink lines from the plate are transferred to the paper.

1.3.2.4 Other intaglio processes
- Etching
- Engraving
- Drypoint
- Aquatint
- Mezzotint
- Photogravure

Engraving and drypoint are processes where the image is scored, scratched or gouged into the surface of the printing plate. Analogous to ploughing a field, a sharp needle creates minute furrows on either side of the line called burrs. During the inking process of drypoint, the ink adheres to the burrs resulting in the appearance of a soft line when printed. Drypoint plates are quite fragile, and the burrs wear out after about 20 prints [13]. Traditionally, drypoint has been done with copper plates, but it can be done on zinc, plastic, and even cardboard. The difference between drypoint and engraving is that the burr is removed for engraving. The ink sits now in the incised line from where it is transferred to damp paper.

The aquatint process involves covering the surface of the copper plate with powdered rosin—an acid-resistant fine resin powder. As the plate is heated, the powder melts and adheres to the surface to create an open ground. The plate is dipped into acid, and the metal is etched around the small particles. Analogous to making a watercolour, hence the term aquatint, individual layers of thin paint resist are applied to increase the tonal range. A stopping-out technique is used to create a range of tones. Working from light to dark, to protect the lighter areas of the image, and obtain a series of deeper tones, further layers of resist are added to the plate surface, and the plate is dipped into acid incrementally until a full range of tones is achieved. The ink is held in the lower parts of the plate; the higher parts are wiped clean.

The mezzotint process is possibly considered the most subtle of printing methods. Printmakers reproducing paintings used mezzotint because of its ability to create chiaroscuro images, smooth gradations in tone, and rich, deep velvet blacks. The mezzotint is a process in which the engraver roughens the entire surface of a copper plate with a very sharp curved cutter with a serrated edge (a 'rocker'). The lines are incised into the surface at many angles, uniformly roughened so that the surface 'ground' has tiny prickly burrs. If the plate is inked at this stage, the print taken from the plate will appear completely black. The engraver then works back into the plate to remove portions of the burr or burnish areas to hold less ink and obtain a range of dark, mid-grey, and light areas. During the inking process, the ink is wiped into the plate and is caught by the roughened surface. Paper is pressed onto the surface and run through an etching press under pressure.

Developed in the eighteenth century by William Henry Fox Talbot, the photo-gravure is a photomechanical intaglio process. It employs a thin paper pre-coated with gelatine and red pigment (previously carbon black) and sensitized with potassium dichromate. After the tissue is applied to a copper plate, it is exposed to a photo-positive image and washed in warm water to remove the more soluble gelatine and develop the image. Once dry, the plate is aquatinted to create a ground, which is then etched using varying strengths of ferric chloride to obtain a tonal range. The plate is inked and printed according to the intaglio method [18].

1.3.2.5 Contemporary methods of intaglio printing and their applications
Flexography or flexo printing is a twentieth century evolution from photogravure and can achieve a high-quality photographic intaglio print (figure 1.15). As well as detail and continuous tone, it can print solid colour. As an industrial process, it can print at vast speeds, thus making it a highly commercial process with applications for printing onto a range of large to small-scale products and packaging from cardboard boxes to milk cartons.

1.3.2.6 Process
The surface of the plate comprises a photosensitive polymer layer on either a metal or Mylar backing that is exposed to UV light through the so-called screen, a pattern of random dots which defines the black in print by producing dimples in the plate. A second exposure through a transparent film carries the positive of the image. The UV light closes the dimples of the random screen, and the whitest white of the print is where all dimples are closed. The plate is then developed in water. The uncured photopolymer is washed away, and after drying and hardening (the plate is flooded with UV to ensure that the photopolymer is cured from the surface to the substrate),

Figure 1.15. Photogravure from a polymer plate. *Der Stiefel*, Susanne Klein, 2020.

the plate is inked, wiped and printed on an etching press. The plate is processed and inked using the same method as etching or gravure.

A significant aspect of flexo printing is the ability to use stochastic halftoning to improve the visual qualities of a photomechanical print. Stochastic technology allows the mechanical simulation of nineteenth century continuous tone. Stochastic is derived from the Greek word *stokastikos*, meaning to predict or guess, and is used in mathematics to analyse and predict the movement of particles through a liquid. In printing, it is a method that applies the a more random-dot frequency modulated (FM) screening. Unlike the traditional halftone system, the process employs no fixed grid or screen angles, and the size of the dots and spacing between the dots is variable. Therefore, more dots can be placed in a specific area to achieve a greater concentration of colour. Also, due to the removal of screen angles (used formerly to prevent moiré effects), the stochastic system enables more than four colours to be printed.

Extra colours have been added to the process colour range, including process orange and green, which has created a greater depth of colour and range of tone, and improved contrast.

1.3.2.7 Rotogravure

At the beginning of the twentieth century, rotogravure became the printing process of choice for newspapers since it enabled the accurate reproduction of artworks and full-colour photographs. Image information is either etched or written directly by laser on a metal printing cylinder, also known as a gravure cylinder. The ink is held in cells which can vary in size and depth, allowing a wide tonal range.

Depending on the application, the press can run at speeds between 300 m min^{-1} and 900 m min^{-1}. Rotogravure is used for newspapers, magazines, speciality printing (printed electronics, for example), and packaging.

1.3.3 Lithography

1.3.3.1 History

Lithography was invented at the end of the eighteenth century by Alois Senefelder of Prague, who could not afford the expensive engraving of plates. He devised a new method by observing the incompatibility of grease and water when applied to a fine grain limestone [19]. Also called 'stone printing' or 'chemical printing', the French term 'lithography' is more widely used.

This process was excellent for colour images and specialist areas such as cartography. Lithographers fought hard against the mechanization of their craft, but by the mid-nineteenth century the power press was introduced, and in the 1870s grained zinc plates replaced the limestone.

1.3.3.2 Definition

Lithography is a planographic method of printing based on the antipathy of oil and water (figure 1.16). A positive image is applied using greasy ink onto a porous limestone block.

Figure 1.16. Full colour metal lithograph. *Dragon Tree*, Susanne Klein, 2020.

1.3.3.3 Process

Drawing materials include a greasy ink called tusche, a wax crayon, a pencil or the grease from one's hand applied onto a limestone block or grained flexible zinc plate. The surface is then etched using a solution of weak nitric acid (HNO_3) and gum arabic. As the surface of the printing plate is rolled with ink, the greasy marks (the positive image) on the plate attract the ink and repel water; the blank etched areas (negative areas) are hydrophilic, attracting water but repelling ink. When printing, the stone or plate is kept damp with water to maintain these conditions. Paper is placed onto the surface, and the stone or plate is run through a press under pressure. The resulting image is a mirror image of the image on the stone or plate.

The industrial version of stone lithography is photo-offset lithography or short offset. In offset printing, the image is not directly transferred from the plate to the substrate but first onto the so-called blanket, a rubber sheet mounted on a cylinder. The rubber cylinder is flexible and allows printing on various materials, including wood, metal, fabric, leather and rough paper or cardboard. The image on the final substrate has the same orientation as on the plate since it has been transferred twice.

The advantage of offset printing is that it produces a consistently high image quality, with quick and easy production of the printing plate and low cost. Modern offset machines are fully automated and allow a fast turnover of print jobs with an hourly production of about 15 000 prints.

1.3.4 Screen printing, silkscreen or serigraphy

1.3.4.1 History

Screen printing was first mentioned in about 221 AD in China [20] and was used to print on fabric. The Japanese used paper or parchment stencils on screens. This technique was introduced to Europe in the late eighteenth century, but the only took off when photosensitive emulsions and thin-film inks were developed in the early twentieth century. In the 1960s, pop artists including Peter Blake, Robert Rauschenberg, and Andy Warhol used screen printing to create highly colourful and visually striking pop art.

1.3.4.2 Definition

The screen printing process involves pushing ink through a stencilled mesh to create an image.

1.3.4.3 Process

The screen printing process involves a fine nylon mesh stretched over a frame and a stencil used to inhibit the transfer of ink through the screen into the paper. A range of materials can be used as the stencil, including paper and hand-painted varnish. The modern version is a photosensitive emulsion, which is uniformly coated across the screen. When the emulsion is dry, the coated screen is exposed to UV radiation using transparent halftone photographic images or drawings on transparent film. After washing the cured image and drying the screen, the frame is placed onto the substrate and ink is forced through the screen with the help of a floodbar or squeegee.

1.3.4.4 Contemporary methods of screen printing and their applications

Since the screen mesh is flexible and not much pressure is required to apply the ink, images can be applied to a wide range of substrates, including paper, textiles, plastic, metal and glass (figure 1.17). It can also be used as a rotary method, for example, onto glass bottles and beer glasses. Screen printing is the main process used for printing on textiles.

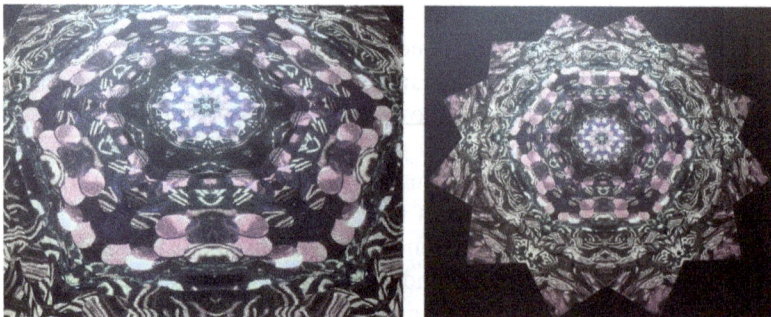

Figure 1.17. Screen print with Spectraval pigments on black paper. (Image by C Parraman.)

1.3.5 Digital printing

1.3.5.1 History

The youngest printing method is the digital print or non-impact print. Invented by Chester Carlson in 1938, the oldest non-impact printing method is xerography. He applied for a US patent in April 1939, which was granted in 1942 [21]. The first commercially successful copier was the Xerox 914, built-in 1959. Ten years later, Gary Starkweather invented the laser printer [22].

Ichiro Endo invented the inkjet printer at Canon [23] and John Vaught at HP [24], simultaneously.

1.3.5.2 Definition

A digital image file is generated and printed directly onto a substrate. As this is a plate-less printing process, the image has to be created anew for every print, which makes the method versatile and slow.

1.3.5.3 Process

1.3.5.3.1 Xerography

A printing drum is electrically charged and partly discharged when exposed to light. The pattern written on the drum is the image. The toner, a triboelectrically charged powder, is brought into contact with the drum and adheres to the drum. The image is then transferred to paper by applying a voltage which counteracts the charge of the drum. The image is fixed by heat and pressure.

1.3.5.3.2 Laser printer

The laser printer is a further development of xerography. The main difference is that a laser writes an image on the electrostatically charged cylinder.

1.3.5.3.3 Inkjet

The image is written directly on the substrate by microscopic ink droplets. Drops are generated on demand, either via heat (thermal inkjet (TIJ)) or a piezoelectric element generates a pressure pulse in the ink reservoir and the ink is expelled (piezoelectric inkjet).

1.3.5.4 Industrial applications

Inkjet is the most common technology for printers in the home and office. Non-impact printing has found industrial applications where customization and low run numbers are key, for example, billboards, banners, lorry sides, catalogues, customized packaging and technical drawings (figure 1.18).

1.4 What is the future of print?

Is print dead? Yes and no. Print, in the sense of distributing information, what it has been since Johannes Gensfleisch zur Laden zum Gutenberg engineered the letterpress in the fifteenth century, is in decline. Over are the days when leaflets called for a revolution or announced the latest bargains in a supermarket. Books have become

Figure 1.18. Assimilation and Contrast, digital print on two layers with laser cut (2024), artwork and image by Carinna Parraman.

collectors pieces again, printed for their aesthetics, but no longer for their information content. Information, and misinformation too, is now easily available on our cyborg extension, the mobile phone, and can be distributed worldwide within milliseconds. No newsroom is working towards printing deadlines anymore. Paperbacks have mostly been replaced by reading apps. COVID-19 has accelerated this trend. According to the Smithers report [25] the drop in the worldwide consumption of print substrates from 256.2 million tonnes in 2019 to 229.6 million tonnes in 2020 reflects a drop in printed material, mostly newspapers and magazines. Packaging was the least affected, especially with the increase of home orders during lockdown. The Smithers reports predicts that by 2030 packaging will account for almost two thirds of the global print market. This is a well-informed prediction, but print is so much more. With the World Wide Web at our fingertips and an almost infinite choice of products, customization has become a tool to make an item special and valuable. Print-on-demand comes in all shapes and on all substrates. Customizing apparel, for example, for sport clubs, businesses or just for a single customer is not new, but it has become so much easier. Many companies offer even single items with customer designed prints, often for not much more than when bought as a mass-produced item. Print is also a very efficient manufacturing method. Electronics are an essential ingredient of almost everything in our daily lives, from the kettle boiling water for our first cup of tea in the morning, to our computer or mobile phone, now essential for work, leisure, and shopping. Printing electronics has been researched heavily for decades without major breakthroughs. That industry and academia still invest in this kind of research shows how much printed electronics is expected to not only reduce the cost of electronics for low performance applications, for example smart labels or the conduction fields of induction hobs, but also to make new applications possible, for example printed sensors on apparel. Packaging itself is a platform for functionalized print. Packaging is not only made attractive by print, but it also carries a lot of information, such as content description, warning labels, use- and sell-by dates, seals, authentication labels etc. Print can change colour when the protective atmosphere of packaging is compromised, or the item was exposed to heat or UV. Labels can certify that the product is genuine. A bar code is not only for scanning at a till but can contain so much more: country of origin, the manufacturer, patient information, web links, booking information etc.

Print is fun too. No student room without posters, no birthday without a birthday card. 2.5D print allows structure and a tactile quality to be given to all kinds of items —a birthday card can become almost 3D, or a shopping bag can be made out of paper but look and feel like it is made of leather or fabric.

We have not included 3D printing in this book since it is a misnomer in our opinion. Printing means that ink and the substrate become inseparable, and this fusion is a print. In 3D printing the 'ink' becomes the item. If substrates are present, they are often discarded when the addition of material is finished. 3D printing is additive manufacturing, and its technical requirements are very different to traditional and non-traditional print. We will concentrate on the multiple substrates where print is possible and how they interact with ink.

Print is dead, long live print.

References

[1] Das R 2019 Printed electronics: the defining trends in 2019 *IDTechEx* https://idtechex.com/en/research-article/printed-electronics-the-defining-trends-in-2019/18340 (Accessed: 25 November 2021)

[2] £10 note *Bank of England* https://bankofengland.co.uk/banknotes/polymer-10-pound-note (Accessed: 25 November 2021)

[3] Peddie R A 1917 *An Outline of the History of Printing: To Which Is Added the History of Printing in Colours* (London: Grafton and Co)

[4] Diderot D and Le Rond d'Alembert J 1751–1722 *Encyclopédie, ou dictionnaire raisonné des sciences, des arts et des métiers (Encyclopedia or Reasoned Dictionary of Sciences, Arts and Crafts)* https://archive.org/details/encyclopdieoudi00conggoog/page/n3/mode/2up (Accessed: 25 September 2023)

[5] Twyman M 1998 *Printing 1770–1970: An Illustrated History of its Development and Uses in England* (revised edn) (London: British Library Publishing)

[6] Parraman C and Ortiz Segovia M V 2018 *2.5D Printing Bridging the Gap Between 2D and 3D Applications* (Hoboken, NJ: Wiley)

[7] Eisenstein E L 1980 *The Printing Press as an Agent of Change: Communications and Cultural Transformations in Early Modern Europe* **vol 1 and 2** (Cambridge: Cambridge University Press)

[8] Briggs A 2005 *A Social History of the Media: from Gutenberg to the Internet* 2nd edn (Cambridge: Polity)

[9] Saff D and Sacilotto D 1978 *Printmaking: History and Process* (Fort Worth, TX: Harcourt Brace)

[10] Martin J 2018 *The Encyclopedia of Printmaking Techniques: A Unique Visual Directory of Printmaking Techniques, with Guidance on How to Use Them* (Tunbridge Wells: Search Press)

[11] Krejca A 1991 *Die Techniken der Graphischen Kunst, Handbuch der Arbeitsvorgänge und der Geschichte der Orginal-Druckgraphik* (Hanau: Werner Dausien)

[12] Covey S 2016 *Modern Printmaking: A Guide to Traditional and Digital Techniques* 1st edn (Berkeley, CA: Watson-Guptill)

[13] Hughes A d A and Vernon-Morris H 2008 *Printmaking: Traditional and Contemporary Techniques* (Mies: RotoVision)

[14] Zapka W 2018 *Handbook of Industrial Inkjet Printing: A Full System Approach* (Weinheim: Wiley)

[15] Meggs P B 2016 *Meggs' History of Graphic Design* (Hoboken, NJ: Wiley)

[16] Kipphan H 2014 *Handbook of Print Media* (Berlin: Springer)

[17] Flexographic Technical Association 2013 *Flexography: Principles and Practices 6.0* (Bohemia, NY: Flexographic Technical Association)

[18] Crawford W 1979 *The Keepers of Light: A History and Working Guide to Early Photographic Processes* (Dobbs Ferry, NY: Morgan and Morgan) p 318

[19] Senefelder A and Lessing J 1818 *Vollständiges Lehrbuch der Steindrukerey ... belegt mit den nöthigen Musterblättern, nebst einer vorangehenben ausführlichen Geschichte dieser Kunst* (München: K Theinemann) Lessing J Rosenwald Collection (Library of Congress) https://www.loc.gov/item/67059021/

[20] Fortune D 2016 *Water-based Screen Printing and Safe Ceramic Decals* (self-published)

[21] Carlson C F 1942 Electrophotography *US Patent Specification* US2297691A

[22] Starkweather G K 1968 Apparatus for forming half-tone line screen with a lens *US Patent Specification* US3535036A

[23] Endo I 1978 Liquid jet recording process and apparatus therefor *Canada Patent Specification* CA1127227A

[24] Donald D K, Lee M J and Vaught J L 1980 Method and apparatus for drop-on-demand inkjet printing *US Patent Specification* US4336544

[25] Smyth S 2020 The future of print to 2030 *Market report* Smithers https://smithers.com/en-gb/services/market-reports/printing/the-future-of-print-to-2030 (Accessed: 9 July 2023)

Chapter 2

Printing on different substrates

Different substrates require different printing methods and different inks. There are three rules:

Rule 1: *Printing hard on hard is not possible.*

When the substrate and the printing plate are stiff the image cannot be transferred successfully. Increasing the pressure will lead to damage to either the substrate or the printing plate. This is the reason why screen printing and offset printing are so important industrially. In both cases the surface, which will come into contact with the substrate, is elastic, can deform and therefore accommodate an uneven and hard substrate. Inkjet is a capable of printing on hard surfaces as well, but the construction of the printer restricts the geometry of the surface it can print on.

Rule 2: *The surface chemistry of the substrate and the ink must be compatible.*

A polar ink will not wet a non-polar surface and vice versa, so if using incompatible materials an image cannot be transferred because the ink will immediately from droplets and the image information will be distorted or even lost. It is easier to print on absorbent surfaces since the ink will penetrate the surface and is then held in place by the pores.

Rule 3: *The ink must stick.*

Even when the ink and the surface are compatible, there is no guarantee that the ink will adhere to the surface. For example, when the solvent content is high and the substrate is non-porous, the image can shrink so much upon drying that it will simply fall off because the forces caused by shrinkage are higher than the adhesion forces.

In this chapter we will discuss the most popular substrates for print. In principal it is possible to print on anything, but to make it a permanent print may require a lot of research and investment.

doi:10.1088/978-0-7503-2568-4ch2
2-1

2.1 Printing on paper

2.1.1 What is paper?

Paper means different things for different industries and different people. We will adopt the definition given in [10]: '[Paper] is a sheet material made up of a network of natural cellulosic fibres which have been deposited from an aqueous suspension'. This definition makes two points: (i) paper is made of cellulosic material and (ii) it is deposited from an aqueous suspension which distinguish paper from felt for example, where the fibre is a protein fibre, and from papyrus, which gave paper its name but is actually not a paper since it is made by layering strips of the stem of the papyrus plant and then hammering or pressing them together until the layers fuse [11].

The paper industry is one of the largest industries in the world. In 2019 the world production of paper and cardboard was 419.69 million metric tons with a global consumption of paper and cardboard per capita of 57 kg and 229 kg in the US for example [12]. More than half of it is packaging.

On one hand the paper industry benefits from the mounting concern about plastic products and plastic packaging, on the other hand traditional paper production generates toxic wastewater which has come under public scrutiny. This problem is increasingly being addressed and paper mills are installing closed-cycle pulp systems [13].

Even though graphics paper production has been falling since 2015, the paper and forest-products industry is growing [14].

Figure 2.1 shows that the strongest growth sector is tissue paper followed by containerboard. Almost all the paper products in figure 2.1 will have some print on them.

Table 2.1 lists the main types of industrial paper and their application. Paper is still the main printing substrate even though the classic application 'book and newsprint' is declining.

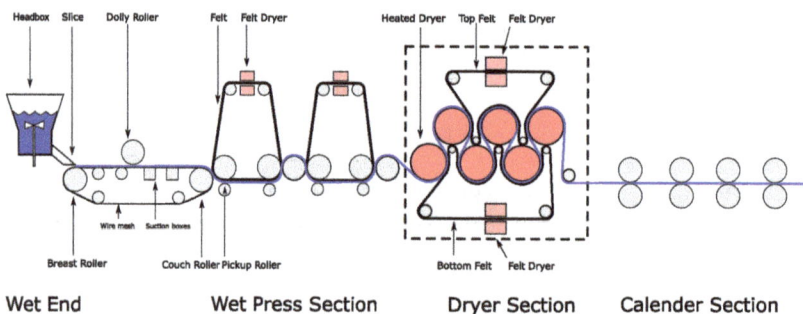

Figure 2.1. Diagram of a Fourdrinier paper machine: paper-making from start to finish. (This image has been obtained by the author from the Wikimedia website where it was made available by Egmason (2010), under a CC BY 3.0 license. It is included within this chapter on that basis. It is attributed to Egmason.)

Table 2.1. Types and applications of industrial paper.

Type	Subtype	Pulp	Pulping process	Application
Printing and writing paper	Uncoated free sheet paper	Mostly wood	Chemical	Office reprographics (copy paper), books, envelope paper and business form paper.
	Uncoated mechanical paper	Mostly wood	Mechanical	Newsprint, newspaper inserts, directories and paperback books.
	Coated free sheet paper	Mostly wood	Chemical	Highly illustrated books, high-quality posters, magazines and advertising pieces such as catalogues.
	Coated mechanical paper	Mostly wood	Mechanical	Magazines, catalogues and coupons.
Containerboard	Single wall	Wood and recovered paper	Chemical	Packaging
	Double wall	Wood and recovered paper	Chemical	Packaging
Paperboard	Bleached paperboard or solid bleached sulphate (SBS)	80% woodchip, 20% recycled. Coated with kaolin clay. If the packaging is intended for frozen foods or liquids, it may also be coated with a very thin plastic lining for wet-strength protection.	Chemical	Packaging food and beverages, cosmetics and pharmaceuticals. Medical packaging. Hot and cold paper cups. Milk and juice gable top cartons. Aseptic drink boxes. Cosmetic and perfume packaging. Frozen food packaging. Candy boxes. Stand up displays.

(*Continued*)

Table 2.1. (*Continued*)

Type	Subtype	Pulp	Pulping process	Application
	Coated unbleached Kraft paperboard	80% woodchip, 20% recycled. Coated with kaolin clay.	Chemical	Packaging food and beverages, cosmetics and pharmaceuticals. Frozen food packaging. Pharmaceutical packaging. Beverage carrying containers.
	Uncoated recycled paperboard	Multi-ply material, produced from recovered paper-collected from paper manufacturing and converting plants and post-industrial sources.	Chemical	Packaging. A top layer of white recovered fibre can be added or the material can be mass dyed to desired colours. Shoeboxes. Composite cans and fibre drums. Coated paperboard.
	Coated recycled paperboard	100% recovered paper. A thin layer of kaolin clay. A layer of white recovered fibre. Manufacturers choose coated recycled paperboard over uncoated when the packaging will be used for products that have marketing and writing on the outside.	Chemical	Soap and laundry detergent packaging. Cookie and cracker packaging. Paper goods packaging (facial tissue and napkins). Cake mix packaging. Cereal boxes. Other dry food packaging.
Tissue	At home or consumer product	Wood, recovered paper	Chemical	Toilet paper and facial tissue, napkins and paper towels, wipes, and other special sanitary papers.
	Specialty tissue papers	Wood, recovered paper	Chemical	Decorative papers that are glazed, unglazed, or creped, and include wrapping tissue for gifts and dry cleaning, as well as crepe paper for decorating.

2.1.1.1 What can paper be made from?

The raw materials for paper production can be classified into three categories: wood, non-wood, and non-plant (non-tree), i.e. waste paper. The majority of paper is made from wood (63% of the world production), followed by wastepaper (34%) and only 3% of the world paper production is based on non-wood pulp [15]. Non-wood paper plays a larger role in Asian countries. In China and India, for example, 70% of the raw fibre for paper production comes from non-wood materials [15] such as bagasse, corn straw, bamboo, reeds, grass, jute, flax, sisal, etc. In Europe and North America high-end paper for hand printing is often cotton based. In the late nineteenth and early twentieth century cotton rags were the main source for paper production and the textile and paper industry grew side by side. When the rag supply could not satisfy the demand of the paper industry anymore, wood became, at least in Europe and North America, the main source of virgin fibre [10].

All paper pulping materials, except cotton and giant reed, have the same main components: cellulose, hemicellulose and lignin (see table 2.2).

2.1.1.2 Cellulose

Cellulose is the primary component in the cell walls of all true plants and the main component of paper. Other types of organisms produce cellulose as well, such as some algae, bacteria, sea squirts (a sea animal) and Oomycetes ('water moulds'). Cellulose seems to have been independently created several times during evolution

Table 2.2. Composition of paper pulp materials from [15, 16].

Raw material	Fibre length (mm)	Fibre diameter (μm)	Cellulose (%)	Hemicellulose (%)	Lignin (%)
Softwood	2.5–4.5	14–65	40–45	20	25–35
Hardwood	0.4–1.8	12–36	40–45	15–35	17–25
Bamboo	2.7–4	15	52–68	15–26	21–31
Reed	1.5–2.5	20	42–50	20–23	22–25
Giant reed	1.2	15	49.8	24–25	—
Bagasse	1.0–1.7	20	55	27–32	18–24
Rice straw	0.5–1.4	8–10	28 -36	23–28	12–16
Wheat straw	1.0–1.5	13	29–35	26–32	16–21
Cornstalk	1.0–1.5	16–20	36–38	23–25	18–19
Kenaf	2.6	20	53	21–23	15–18
Flax	25–30	20–22	70	6–17	10–25
Hemp	20	22	57–77	9–14	5–9
Jute	2.0–2.5	20	57	15 -26	16–26
Cotton	20–30	20	95–97	—	—
Cotton linter	0.6–3.0	20	90–91	—	3
Sisal	3.0–3.5	17–20	43–56	21–24	8–9
Abaca	6.0	20–24	61	11	9

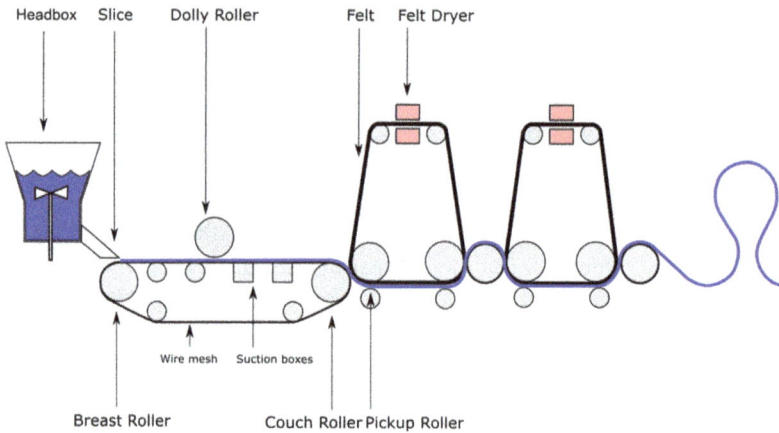

Headbox   Slice   Dolly Roller   Felt   Felt Dryer

Wire mesh   Suction boxes

Breast Roller   Couch Roller Pickup Roller

Wet End

Figure 2.2. The wet end of the Fourdrinier paper-making process. (This image has been obtained by the author from the Wikimedia website where it was made available by Egmason (2010), under a CC BY 3.0 license. It is included within this chapter on that basis. It is attributed to Egmason.)

but always has the same covalent pattern. It is always a mechanical enforcing fibre. It can occur either pure (in cotton) or more commonly mixed with lignin and other polysaccharides. Cellulose is a very strong material: cellulose fibrils are stronger than steel of the same dimensions [16]. It is totally water insoluble, but its surface is very hydrophilic. Cotton for example can absorb ten times its weight in water. Even though the surface of cellulose is hydrophilic, it interacts well with hydrocarbons since the top and bottom of the cellulose sheets are rather hydrophobic (figure 2.2).

It is a very simple unbranched polymer of polysaccharide of β-1,4 linked D-glucopyranose, see figure 2.2(a). The degree of polymerization can be very high, up to 15 000 glucopyranoside residues [16], which makes it one of the longest known polysaccharides. Wood cellulose does not have a particularly high degree of polymerization. The highest molecular weights, a measure of the degree of polymerization, is found in non-wood sources such as flax and cotton. The secondary structure of the cellulose decides whether it will be of any industrial interest. Hydrogen bonds between chains make the structure stiff and arrange them into sheets (figure 2.2(b)). The sheets stack on top of each other and interact via van der Waals forces forming two different crystal structures I_α and I_β. I_α is metastable and can be transformed into I_β by high temperature and pressure in an alkaline or acidic solution [16]. The sheets themselves are narrow and therefore do not form a continuous solid crystal but bundle into so-called fibrils which vary dependent on where they occur, for example they are different in leaves and wood. A cellulose chain can be 5–7 μm long but its superstructure, the fibril, can reach lengths of up to 40 μm (figure 2.3). Cellulose crystalizing into I_α and I_β is called cellulose I and is the naturally occurring variety. When treated chemically it can

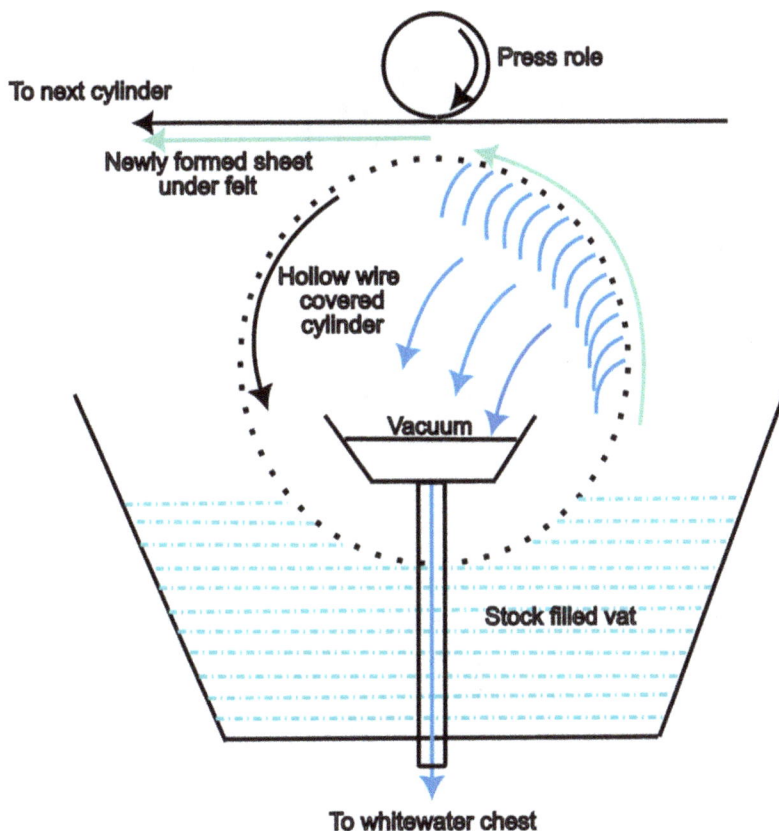

Figure 2.3. The cylinder mould machine invented by John Dickinson in 1809.

rearrange into different configurations called cellulose II, III and IV. Cellulose II is industrially the most important and is the basis of rayon fibres, cellophane and Lyocell fibres.

Hydrogen bonding is the mechanism which makes the fibres stick to each other. Since the hydrogen bond length is only of the order of a few nanometres, the two fibre surfaces have to come into very close proximity. Pulp is a suspension of cellulose fibre in water with a fibre content between 0.1 and 4 wt% [16]. During drying the fibres in the pulp are forced together.

How the paper dries has a large influence on the final properties of the paper. Industrially manufactured paper is dried by contact drying with heated cylinders (figure 2.4). In the outflow of the headbox, the slurry is air-free and has a water content between 99.9 and 96 wt%. In the wire section the water content is decrease to 80 wt%. In the press section it decreases further to 60%. Proper drying happens in the drying section. Water evaporates from the paper mass between the steam-heated rollers. After the drying section the web has lost almost all its water. Hydrogen bonding sets in during the press section. The swollen fibres are surrounded by a water film. Between the fibres the water forms a meniscus. The smaller the radius of

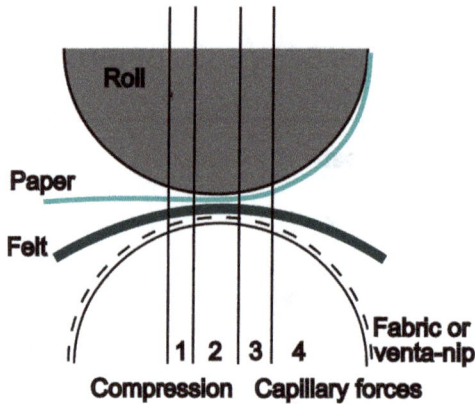

Figure 2.4. The four stages in the nipping process.

the meniscus, the stronger the attractive force between the fibres. The more water evaporates the more the fibres are pulled together until they are so close that OH and NH$_2$ groups on the surface of the fibrils can form hydrogen bonds. Hydrogen bonds form when the molecules are less than 0.5 nm apart. Attractive van der Waals or London dispersion forces become important much earlier, at about 10 nm, and increase the attractive pull between the fibres even more. Water between the fibres and its increasing pulling force during drying gives the dry paper its dry-strength. Frieze dried sheets have a much lower dry-strength then heat-dried sheets [16] since the water is frozen and then removed by sublimation, i.e. the fibres are never pulled together. The softer the fibres the easier it is to pull them into close contact. Chemically treated pulp contains well beaten and swollen fibres which allow good contact between surfaces, especially since lignin has been removed by the chemical treatment. Mechanical pulp contains relatively stiff fibres which are partly covered in lignin, decreasing the contact area for hydrogen bonding and having less overall area with strong van der Waals interactions. Mechanical pulp leads therefore to less strong dry paper and cardboard [16].

Drying of paper has not only nano- or microscopic effects, but also macroscopic ones: shrinkage, cockling and curl. One problematic dimensional instability of paper products during printing is the out-of-plane curl when water-based inks are used. A moisture gradient develops through the whole thickness of the paper layer and leads to a distortion called curl or cockling upon drying which occurs as well when paper is exposed to moisture after printing (see figure 2.5).

All dimensional changes of a sheet of paper have their origin in free and bonded fibre segments within the fibre network of the sheet. Their distribution, orientation, and characteristics determine whether a sheet expands or contracts upon changes in moisture content. The network bonds form at a water content of about 50%. Further reduction of moisture does not change the location of the bonds but leads to significant shrinkage of the fibre when moisture leaves the cell walls [19]. Studies in the 1960s and 1990s [20–22] showed that during free-drying, as the sheet is not clamped or compressed during drying, the bond segment contracts and causes

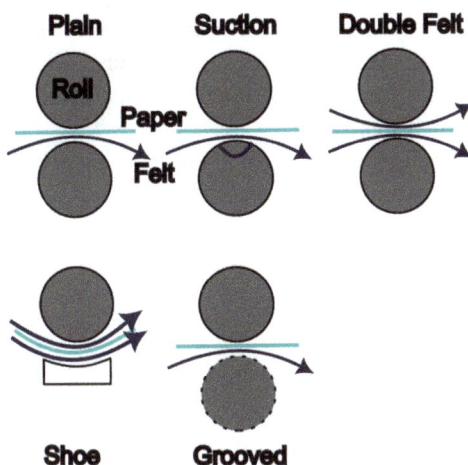

Figure 2.5. Methods to prevent water from returning to the paper in the nip press.

shrinkage of the whole sheet. The fibre walls are anisotropic, and the fibres shrink more in their transversal direction than in their longitudinal direction. The fibres are already bonded and cannot move anymore. At a bond, the transversal shrinkage of one fibre puts a microcompression on the longitudinal direction of the crossing fibre which causes every fibre in the sheet to shorten lengthwise [23]. During restrained drying, the wet sheet is either clamped down or compressed, the contraction of the bond segment causes stretching of the fibre segments between the bond sites and the whole sheet shows none or very little shrinkage.

The strength of wet paper is always much lower than that of dry paper. When dry paper is ripped, fibres have to be separated mechanically, i.e. van der Waals forces have to be overcome and hydrogen bonds have to be broken. The dry-strength of paper is directly proportional to the number of network bonds. When the paper is wet, water molecules diffuse in between the fibres since there is a concentration gradient between the surrounding water and the lower number of water molecules surrounding the dry fibres. The penetrating water molecules bond preferably with the hydrogen molecules of the cellulose, therefore break the bonds between the cellulose molecules and by doing so push the fibres apart lowering the van der Waals forces as well. Many paper and cardboard products need wet-strength. Wet-strength is usually expressed as percentage of dry-strength. When the wet-strength is at least 10%–15% of the dry-strength, the paper is considered as wet-strength paper [16]. Paper which loses its wet-strength gradually over a couple of hours has temporary wet-strength, whereas paper which keeps its wet-strength over longer periods has permanent wet-strength. In the 1930s and 1940s the first wet-strength chemicals were developed. Wet-strength chemicals are mostly applied in the production of hygiene products, such as hand towels, serviettes, toilet tissue, facial tissue and cleaning cloths [24] and give temporary wet-strength. Another area where wet-strength is required is packaging: paper sacks, carrier bags, milk cartons, freeze packaging, meat wrappers and fruit trays need at least temporary wet-strength. Speciality areas are filter paper, wall paper, abrasive paper and labels [24]. Wet-strength chemicals prevent fibre swelling and therefore protect existing bonds and/or form covalent bonds between the paper fibres [16].

Table 2.3 lists commercial wet-strength chemicals, their most likely strengthening mechanism, and their advantages and disadvantages.

Table 2.3. List of commercial wet-strength chemicals, compiled from [16].

Name	Mechanism	Advantages	Disadvantages
Urea-formaldehyde	Protection of bonds	• Low cost • Easy repulping • Permanent wet-strength • No organochlorine or organic halides	• Free formaldehyde
Melamine-formaldehyde	Protection of bonds	• Relatively cheap • Permanent wet-strength	• Free formaldehyde • Has a pH of 2 • Corrosive • An acid-resistant handling system is required
Alkaline-curing resins	• Homo-crosslinking • Co-crosslinking • Reaction with water	• Neutral or slightly alkaline conditions • Permanent wet-strength • High relative wet-strength	• Difficult to repulp • Contains organic chlorine • Reduced absorption • Capacity of paper
Glyoxalated polyacrylamide resin	• Hydrogen bonds with cellulose • Ionic bonds • Homo-crosslinking • Co-crosslinking	• Easy repulping • High absorption capacity of the treated paper • Dry-strength	• Temporary wet-strength • Decomposition of paper at alkaline contact
Starch	• Hydrogen bonds with cellulose	• Easy repulping • Biologically degradable • Dry-strength • From renewable resources	• Less effective than the other wet-strength chemicals • Temporary wet-strength • Decomposition of paper at alkaline contact

2.1.1.3 Hemicellulose

Hemicellulose is the name of a group of low molecular weight, heterogenous polysaccharides. E Schulze (1891) proposed the name for this group of non-structural polysaccharides under the assumption that they are more easily hydro-lyzed than cellulose and were intermediates in the biosynthesis of cellulose [16]. Hemicellulose are unrelated to cellulose and are formed biosynthetically by a separate route [10]. The degree of polymerization is low, between 150 to 200 (cellulose has a degree of polymerization up to 15 000). The main building blocks are of the hexoses D-glycopyranose, D-mannopyranose ad D-galactopyranose and/or of the pentose D-xylopyranose and L-arabinose group (see figure 2.6). Small amounts of the deoxyhexoses L-rhamnose and L-fucose, of 4-0-methyl-D-glucuronic acid, D-galacturonic acid and D-glucuronic acid are present too and the mono-saccharides are often partly acetylated. The chemical and thermal stability of hemi-cellulose is lower than that of cellulose [10, 16]. The physical state of hemicellulose and its function in wood are unclear. It could contribute to the stability of the cell walls and to the capacity of the cell to store and transport water [10, 16]. The major hemicellulose group in softwood is galactoglucomannan and in hardwood it is glucuronoxylan. Table 2.4 lists the major hemicelluloses in soft- and hardwood, but these are not only found in woods but also in grasses, cereals and some primitive plants.

Hemicellulose plays an important role in paper-making. It can improve pulp beatability and acts as an inter-fibre binding agent, improving the tensile, tear and burst strength of the finished paper product [26–29]. Paper sheets where hemi-cellulose has been extracted are weak. Hemicellulose is more hydrophilic than cellulose and enhances swelling of the cell wall leading to increased fibre flexibility which then increases the inter-fibre bonding during network formation.

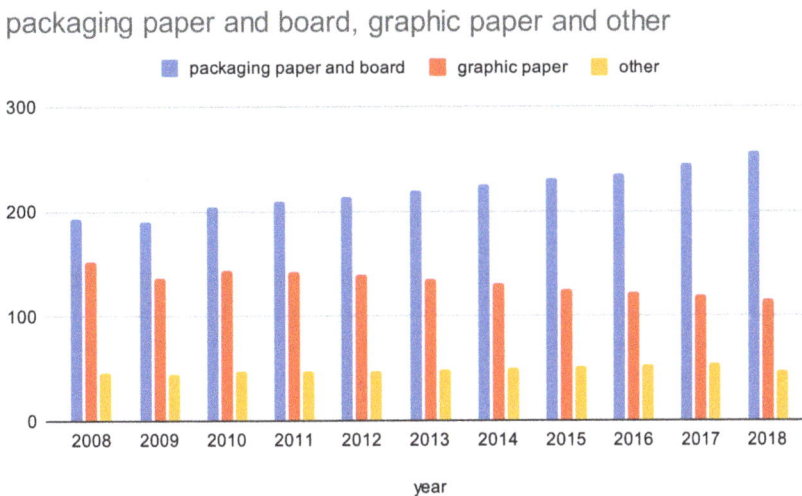

Figure 2.6. Development of the global paper and board market from 2008 to 2018 in million metric tons. (Data from [12].)

Table 2.4. Major hemicelluloses in soft and hardwoods [16].

Occurrence	Hemicellulose	Amount in wt% of dry weight
Softwood	Galactoglucomannan	5–8
Softwood	Glucomannan	10–15
Softwood	Arabinoglucuronoxylan	7–15
Larch wood	Arabinogalactan	3–35
Hardwood	Glucoronoxylan	15–35
Hardwood	Glucomannan	2–5

Pulps with short fibres and recycled paper benefit from the addition of hemi-cellulose. Brightness and opacity increase with a decreasing hemicellulose content but the ink transfer during printing is slightly reduced. Print-through decreases as well with the removal of hemicellulose. On the other hand print density is decreased when hemicellulose is lost or removed from the pulp [26]. It is therefore important to control the hemicellulose content during paper-making.

2.1.1.4 Lignin
Cotton and wood are source materials for the cellulose in paper. They both have high tensile strength but are otherwise very different. Cotton is soft and can absorb water ten times its weight whereas wood is stiff with low water absorption. The chemical component which causes this difference is lignin. Cotton does not contain lignin, but wood contains 17–35 wt% (see table 2.2). It is a strengthening agent in the wood structure, i.e. in combination with cellulose and hemicellulose it makes the fibres stiff, it keeps cells connected (a function fulfilled by pectin in many herbs and non-woody tissue of the tree), it makes the cell wall hydrophobic by inhibiting swelling of the cell walls in water, and protects the wood from micro-organisms and decay [10, 16]. Lignin is one of the most abundant biopolymers, but does not fall into the most common classes of biomolecules. It is neither a polysaccharide, a lipid, a protein nor a nucleotide. It is a very complex mixture of aromatic and aliphatic moieties and forms three-dimensional web. The main monomers of this web, the so-called monolignols, are p-coumaryl alcohol, conifer alcohol and sinapyl alcohol (see figure 2.7).
The three main types of lignin are [16]:
- Softwood lignin or guaiacyl lignin which consists almost exclusively of coniferyl alcohol with traces of p-coumaryl alcohol (see figure 2.8).
- Hardwood lignin or syringyl-guaiacyl lignin which contains coniferyl and synapyl alcohols.
- Grass lignin or HGS-ligning (hydroxyl phenol, guaiacyl, syringyl) which consists of all three types of monolignols and with the highest p-coumaryl alcohol content of all types of lignin.

For the production of high-quality paper with good dry- and wet-strength, the removal of lignin is necessary. Paper products, where no longevity is required, are

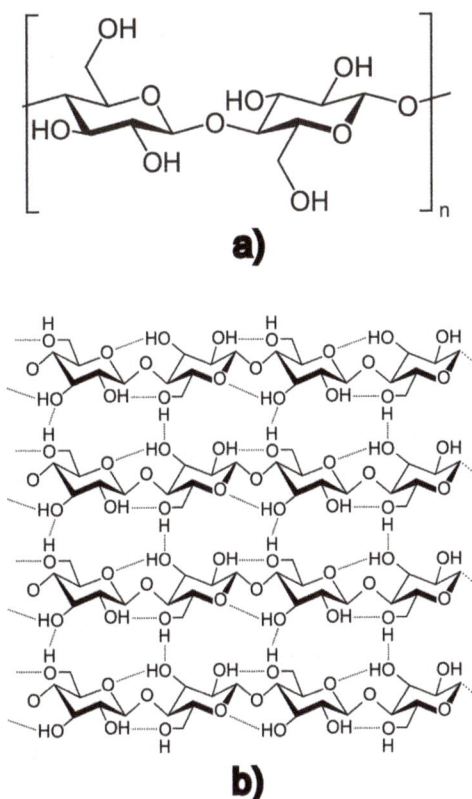

Figure 2.7. (a) β-1,4 linked D-glucopyranose, (b) cellulose sheet connected via H bonds. ((a) This image has been obtained by the author from the Wikimedia website where it was made available by NEUROtiker (2007). It is stated to be in the public domain. It is included within this chapter on that basis. It is attributed to NEUROtiker. (b) This image has been obtained by the author from the Wikimedia website where it was made available by Laghi (2013), under a CC BY-SA 3.0 licence. It is included within this chapter on that basis. It is attributed to Laghi.)

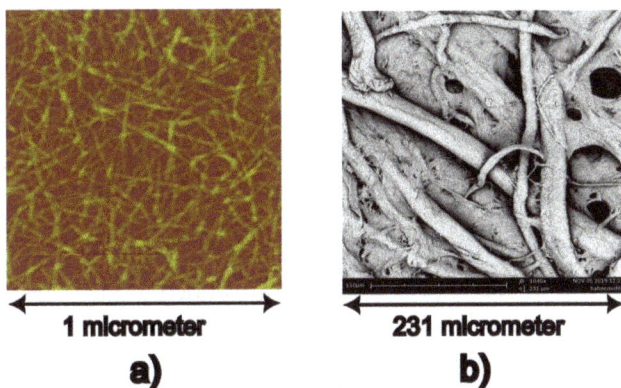

Figure 2.8. (a) AFM height image of carboxymethylated nanocellulose adsorbed on a silica surface. (b) SEM of pure cotton paper. ((a) This image has been obtained by the author from the Wikimedia website where it was made available by Innventia (2010). It is stated to be in the public domain. It is included within this chapter on that basis. It is attributed to Innventia.)

made by mechanical pulping where the lignin is not removed. These paper products will yellow and become brittle because the lignin will photo-oxidize. The dry- and wet-strength of paper with a high lignin content, up to 30 wt%, is compromised. The hydrophobic and stiffening properties of lignin prevent the formation of a strong network. The removal of lignin from the pulp requires a harsh chemical treatment, is costly and the lignin content of wastewater from papermills is environmentally problematic. On the other hand, lignin could be an alternative to petroleum for the production of plastic products or fuels. During the last 15 years, lignin recovery has become of scientific and economical interest, see for example [30–35].

2.1.1.5 Sizing and coating

When the dry paper comes off the paper machine it is rough and inhomogeneous. The surface quality is insufficient for any printing application. For printing and writing paper needs to be 'sized' or coated. When paper or cardboard is sized or coated it can withstand wetting or liquid penetration and the surface strength is increased. Withstanding wetting or liquid penetration is not the same as wet-strength as discussed before. Traditionally paper was treated, 'sized', to stop ink from feathering or print from bleeding. Sizing can mean two things: control of water penetration into the bulk of the sheet of paper or board (internal sizing) or control of water penetration through the surface of the sheet of paper or board (surface sizing). Additionally, in traditional letterpress and lithographic offset printing, the paper surface is exposed to considerable forces which can tear fibres from the surface which then collect on the plate or blanket (picking). The print quality deteriorates, printing must be stopped and the plate/blanket has to be cleaned. Sizing and coating are measures which ensure that the strength of the paper meets the requirements of the printing process. Common for all printing processes is that the highest printing quality can only be achieved when the paper surface is homogeneous in porosity, chemical composition, topography, and reflectance.

There are three ways to improve the surface quality of paper for printing:
- Internal sizing
- Surface sizing
- Coating

2.1.1.6 Internal sizing

Internal sizing prevents liquids from penetrating the paper matrix by treating the fibres in the pulp with hydrophobic substances. It does not prevent vapour from penetrating the paper matrix. For that a physical barrier has to be created which can be done by coating.

Internal sizing is a wet end operation. Sizes are added to the fibre-filler suspension as emulsions of waxy materials with droplet sizes of 1 μm. They should stay in the wet web during sheet formation without hindering the bond formation between the different fibres and migrate during pressing and drying to the non-bonded fibre surfaces in the web [10]. Retention of the hydrophobic sizing agents at the wet end is achieved via ionic interaction. The sizing emulsion is normally cationic and the fibre surface anionic. Different to quaternary ammonium salts, which are used as fibre softener, the sizing molecules do not absorb with the cationic end to the fibre surface at this stage. Only when the water content of the pulp has decreased so much that bond formation between the paper fibres sets in, the sizing molecules adsorb to the fibre surface via strong electrostatic interaction or covalently bond [10].

The oldest method for internal sizing involves the use of wood resin acids (rosin) and aluminium sulphate (alum). Rosin is an extract from softwoods. There are three types:

- Gum rosin, harvested by tapping living pine trees.
- Wood rosin, extracted from aged wood stumps.
- Tall rosin, distilled from the crude tall oil from the kraft pulping process.

Rosins for sizing are used either in free acid form (acid rosin particles dispersions) or as sodium carboxylate salt (saponified rosin size). All rosin sizes are used together with alum.

Sizing systems can be divided in four components: the sizing agent (rosin for example), the precipitator which brings the sizing agent onto the fibre, the substrate (cellulose) and the 'catcher' which orients the sizing agent to the fibre in such a way that the fibre becomes hydrophobic [36]. Cellulose is a substrate and catcher at the same time since its surface is anionic. Alum is precipitator and catcher in the rosin/ alum sizing system by forming an aluminium ester bridge between the cellulose and rosin, see figure 2.9. This is the accepted mechanism for the anchoring of acid rosin dispersions to the paper fibres.

Saponified rosin is precipitated with alum and an emulsion forms with rosin particle sizes between 0.01 and 0.1 μm. These colloidal particles are cationic in an pH interval between 4 and 6.5 if there is a surplus of aluminium ions and adsorb to the anionic cellulose fibres. Rosin/alum sizing needs an acidic pH to work.

Modern paper milling lends itself more to neutral pH sizing systems where the size molecules bond covalently to the hydroxy groups of the fibre surface. Covalent bonds are more robust than ionic bonds, attach the sizes more permanently to the paper fibres and can therefore be used at much lower levels [10]. The commercially

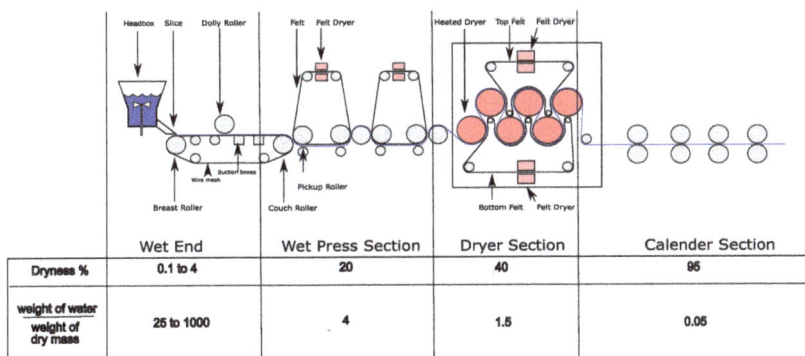

	Wet End	Wet Press Section	Dryer Section	Calender Section
Dryness %	0.1 to 4	20	40	95
weight of water / weight of dry mass	25 to 1000	4	1.5	0.05

Figure 2.9. Industrial paper dryness and moisture content based on [16, 18]. (This image has been obtained and adapted by the author from the Wikimedia website where it was made available by Egmason (2010), under a CC BY 3.0 license. It is included within this chapter on that basis. It is attributed to Egmason.)

2-15

Figure 2.10. Example of curl and cockling of a Woodburytype.

most important neutral sizes are the alkyl ketene dimers (AKD) and the alkenyl succinic anydrides (ASA). They work in a pH range of 5–8.5. The AKDs were developed in the 1940s. They are synthesized from long fatty acids via their acid chlorides which then form the alkyl ketene dimers. The saturated AKDs are waxy substances, with melting points of about 50 °C.

The alkyl chains in figure 2.10 can be between C-14 to C-18. The sizing becomes more efficient with increasing chain length but levels off at C-20 [16].

The ASAs were developed in the 1970s and have a much higher cellulose reactivity than then the AKDs, but are also more prone to hydrolysis, that means that they can be detached from the cellulose by water molecules. The ASAs are water-insoluble oily liquids at room temperature. Their high reactivity makes it possible to achieve most of the hydrophobizing effect before the size press. It is not completely clear how ASA attaches to the cellulose surface, but the most likely mechanism is an ester bond to the OH group on the fibre [37] (see figure 2.11).

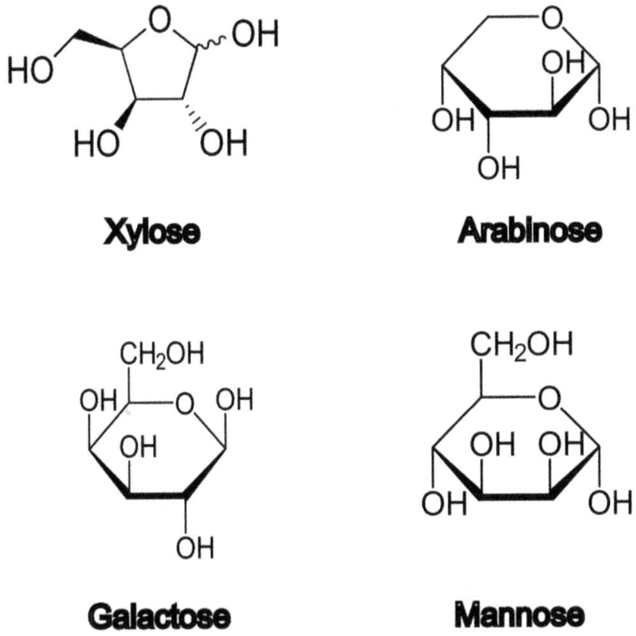

Figure 2.11. Examples of the main components of hemicellulose.

ASA sizing is mostly used in fine paper-making. AKD and ASA sized papers loose hydrophobicity with time, an effect that is not completely understood [10].

2.1.1.7 Surface sizing and coating
Both surface sizing and coating happen at the dry end of the paper-making process. Surface sizing is part of the paper machine operation whereas coating is done on a separate machine. Surface sizing does not hide the fibre structure of the paper surface, but coating evens the roughness by filling the topology of the fibre network. Surface sizing improves the quality of the paper surface and can be followed by coating.

Surface sizes are water-soluble polymers. The four most important ones are starch, carboxymethyl-cellulose, polyvinyl alcohol and alginates [10].

Starch is a polysaccharide, produced by most green plants as energy storage and commercially extracted from potato, maize, rice, cassava and emmer wheat. Pure starch is a white powder and is insoluble in cold water or alcohol. It contains 20–25 wt% of amylose and 75–80 wt% of amylopectin [39], see figure 2.12.

Starch is of commercial interest since it is so cheap, but it can cause problems when paper is recycled since it detaches easily from the fibre and pollutes the wastewater.

As a primer for coating, carboxymethyl cellulose is preferred. It is a cellulose derivative with its structural formula as shown in figure 2.13.

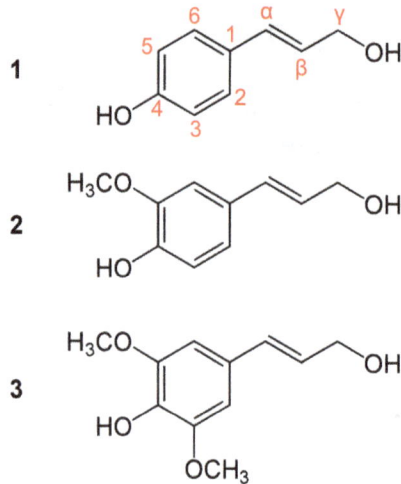

Figure 2.12. The three common monolignols of lignin: (1) p-coumaryl alcohol, (2) coniferyl alcohol, and (3) sinapyl alcohol. (This image has been obtained by the author from the Wikimedia website where it was made available by Yikrazuul (2009). It is stated to be in the public domain. It is included within this chapter on that basis. It is attributed to Yikrazuul.)

When a clear, tough and well sealed film with a high density is desired then alginates are used as the surface size. Alginates are polysaccharides from the brown algal seaweed (figure 2.14).

When a paper with high tensile strength and good oil resistance is required, polyvinyl alcohol (figure 2.15) is the most important surface size. Polyvinyl alcohol is resistant against almost all solvents.

Coated paper and board are printing substrates for lithographic offset, rotogravure and flexography. In artist printing they are used for Woodburytype, relief print and screen printing. The coating layer is applied as a high pigment content suspension (60 wt%) in a binder. In Europe 72% of the pigment used for paper coating is ground calcium carbonate, 20% kaolin clay, and 8% other pigments, such as talc and titanium dioxide for example. In the US mostly kaolin clay (73%) is used, followed by ground calcium carbonate (15%) and 12% other pigments [16]. The particle size of the white pigments is around 2 μm. The binders fall into two categories: latex binders and water-soluble binders. The function of the latex binder is to bind the pigments to each other and the substrate. In addition to that the water-soluble binder acts as a thickener and a water retention agent.

Ground calcium carbonate is made from ground up chalk, limestone or marble. The main mineral of all three rocks is calcite, $CaCO_3$. Chalk is the soft, white and porous sedimentary form, limestone is a harder version of chalk and contains variable amounts of silica. Marble is a metaphoric rock where the calcite recrystallized under high temperature and pressure. Ground calcium carbonate has a much higher brightness than the kaolin clay but has a lower scattering coefficient and does not give a very good coverage. Ground calcium carbonate is used in coatings where the coat weight is high, and a matt appearance is desirable. It is not used as a coating for paper or board for rotogravure [16].

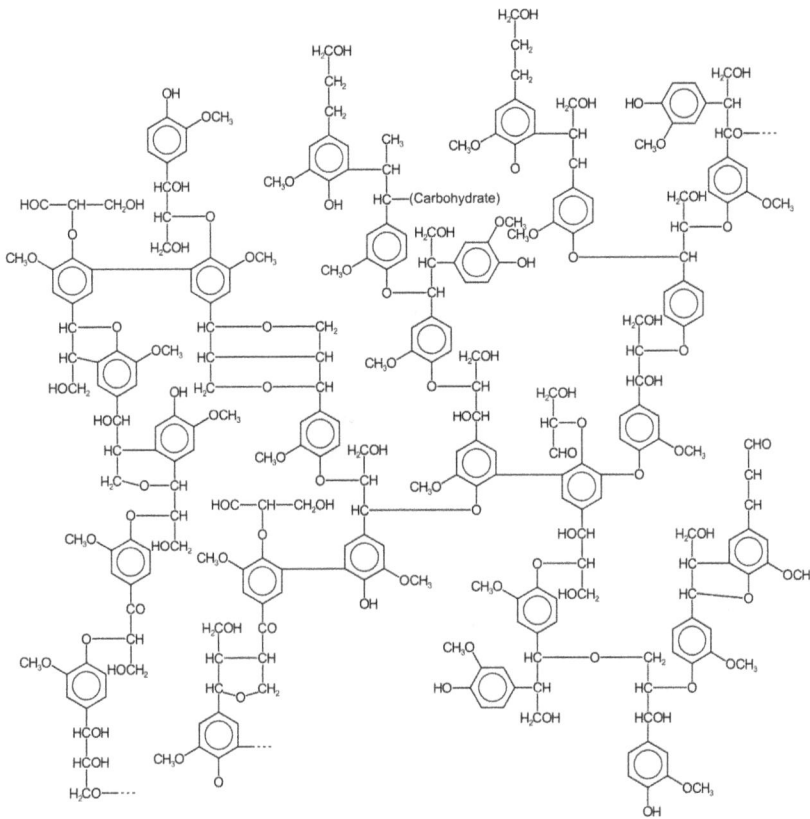

Figure 2.13. An example of a possible softwood lignin structure. (This image has been obtained by the author from the Wikimedia website where it was made available by Karol007 (2007), under a CC BY-SA 3.0 license. It is included within this chapter on that basis. It is attributed to Karol007.)

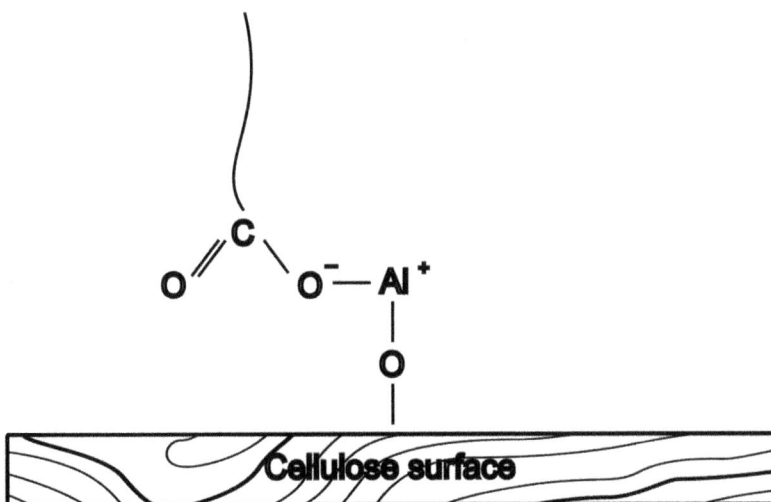

Figure 2.14. Aluminium ester bridge between rosin and cellulose formed by alum.

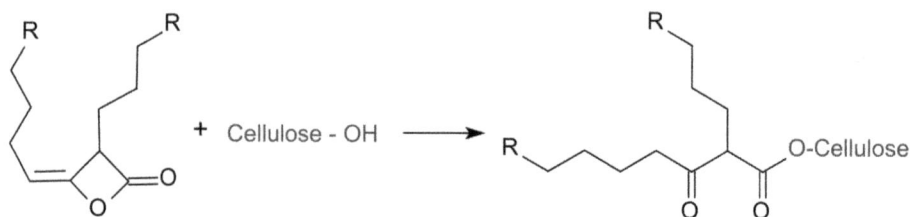

Figure 2.15. AKD esterification with cellulose. (This image has been obtained by the author from the Wikimedia website where it was made available by ChemDoc 2010 (2017), under a CC BY-SA 4.0 license. It is included within this chapter on that basis. It is attributed to ChemDoc 2010.)

Kaolin clay is a weathering product of feldspar. Its main mineral is kaolinite, a layered silicate (phyllosilicate) with the chemical formula $Al_2O_3 \cdot 2SiO_2 \cdot 2H_2O$. It is also known as China clay since it was first reported in Europe as an ingredient of Jingdezhen porcelain. Kaolin coatings excel when it comes to coverage and are therefore used when a light coating is desired. The high aspect ratio of the clay particles and there flat alignment in the coating give good printing properties in rotogravure [16].

The two other pigments mentioned above are talc and titanium dioxide. Talc is a phyllosilicate as well and used mainly for lightweight coated paper for rotogravure. Titanium dioxide comes as rutile or anatase, two different crystal forms of TiO_2. It provides good coverage and is mainly used for low brightness substrates such as like white lined chip board or liquid board.

2.1.1.8 Binders

The latex binders in coatings are monodispersed synthetic polymer dispersions with spherical particles of diameters between 0.15 and 0.2 μm. Latex in coatings comes in three versions: styrene butadiene, a synthetic rubber also used for tyre production, styrene butyl acrylate, also used in architectural and car coatings, and polyvinyl acetate, or wood glue in its non-emulsified version. Amorphous materials, including polymers, display a change in physical properties, from hard and brittle to viscous or rubbery when they are heated to the so-called glass transition temperature T_g. Film formation of coatings with latex as a binder happens during drying when T_g is reached. When water is driven out of the dispersion, pigments and latex particles will aggregate and a film is formed by heating the layer above the so-called film formation temperature which is roughly the glass transition temperature. The latex particles then 'melt' together and form a continuous film encapsulation of the pigments in the coating. T_g affects the stiffness, porosity, light scattering and gloss of the film. When a soft, compressible, and dense coating is desired, for rotogravure for example, a latex with a low T_g is chosen (table 2.5).

A higher T_g means that the latex particles will stick, but not necessarily flow into each other. The porosity of a high T_g film is high and the pigments in the coating are not so well embedded in the coating. Polyvinyl acetate coatings are rather porous because of its high T_g. It not very common in Europe but used widely for board coating in the US.

Table 2.5. Glass transition temperature for latex in paper coating, compiled from [16]. The range of T_g for styrene butadiene and styrene acrylate is a function of the ratio of styrene and butadiene and styrene and butyl acrylate.

Latex	Styrene butadiene	Styrene butyl acrylate	Polyvinyl acetate
T_g	−20 °C to 50 °C	−10 °C to 40 °C	30 °C

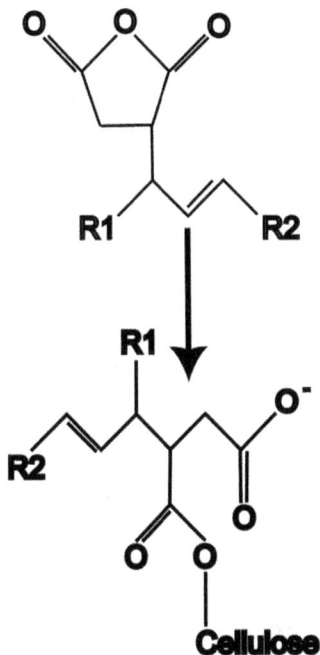

Figure 2.16. Attachment of ASA to cellulose according to [38].

Starch, see figure 2.12, is not only popular as a surface size but also as a water-soluble binder for coatings. Starch for coatings comes from corn and potatoes.

Sodium carboxymethyl cellulose is another common water-soluble, all-purpose binder. It is built from similar monomers as starch but alternating positions of the bonds between the monomers make it an extended molecule and not a coiled one as starch, see figure 2.16.

Coatings are basically pigmented inks, only applied in a different way. As in inks, additional chemicals are part of coatings: dispersants, biocides, hardeners, lubricants and optical brighteners.

Paper and paper products are still the most common substrates for printing. All printing methods in section 1.3 have been developed for paper. How and what is printed on it is dictated by the application and local regulations. The printing process and the ink used will differ radically between for example food packaging, money, or sensors. Since printing on paper is so widespread, there are commercially available inks for almost all applications. The paper substrate itself can present many different surfaces, from very porous and soft, tissue paper, to sealed and hard, coated cartons for packaging. In general print on paper can be recycled but how is again determined by the application, sizing, surface treatment, inks used and how contaminated the substrate is by its use. It is certainly one of the more sustainable substrates because it can be regrown, but paper does not guarantee easy recycling.

2.1.2 The paper-making process

2.1.2.1 General classification of paper
Before making the paper, materials need to be prepared as a pulp. The aim of the pulping process is to break down the bulk structure of the plant into fibres.

2.1.2.2 Comparing mechanical and chemical pulping methods
Groundwood, as it suggests, is wood that is ground mechanically to make a pulp. The grinding process generates heat and softens the wood to enable the fibres to separate, resulting in a short fibre pulp. Very little material is lost in this process, 95% of the woody material is used, but still comprises a quantity of lignin which is highly acidic. If paper products are required to last, the acidic content is unwanted, for example a newspaper over time tends to become brittle and unstable, and oxidizes and yellows in sunlight.

By comparison, chemical pulping methods, such as heating and bleaching methods, are used to chemically dissolve the lignin and soften the fibres, resulting in a whiter and more stable pulp. However, the increase in energy required and the process of removing lignin makes production more expensive, and the yield is typically around 50% less than the mechanical method. For paper products that require greater longevity, for example legal documents, chemical production is the preferred method.

2.1.2.3 From tree to sheet: a brief overview of making paper

2.1.2.3.1 Preparing wood fibre
Wood fibre is shipped from sawmills, and the contents are sorted into large and small chips.

Woodchip and sawdust are used for paper-making, and the bark is burnt to provide energy. The chips are washed and sent to the digester. In their raw state, chips currently contain 50% water, 25% cellulose fibre, and 25% lignin.

2.1.2.3.2 Cooking the fibres
The first step is the cooking process; the aim is to break down the fibres and remove the lignin, which is the fibre's natural glue. During the cooking process, materials are put into a continuous digester. Chips are combined with a liquid, called white liquor, which is a compound of caustic soda and sodium sulphide. By the end of the process, 99% of the chemicals are recovered.

2.1.2.3.3 Breaking down the fibres

In the digester, temperature and pressure break down the fibres to dissolve and separate the pulp. Air is pressed out using low pressure and steaming. Under high pressure of up to 14 bars, the chips are thoroughly impregnated with the cooking liquor. The resulting broken-down material is a mixture of black liquor and pulp. The black liquor is comprised of white liquor and lignin.

The black liquor is sent to the recovery circuit, and the organic matter is burnt to heat water, which produces steam and generates electricity to power the plant. The burnt black liquor leaves a residue, that when mixed with water is called green liquor. Lime is added and it is converted back to white liquor. This residue is used once again as a raw material.

2.1.2.3.4 Removing the water

Any remaining lignin is washed and bleached using chlorine dioxide. Water is removed by squeezing, drying, and draining. Dry pulp sheets are packed in bales and shipped around the world. A lot of water is used as part of the process but is continuously recycled. All water is treated through an effluent treatment process and returned to the environment with no negative impact. The sludge that is generated provides an ideal environment for the growth of micro-organisms.

2.1.2.3.5 Preparing the fibres

The fibre preparation is the first step in ensuring uniform and consistent paper. This includes cutting the fibre to achieve a high fibre concentration and even fibre formation across the sheet and to improve the overall appearance of the sheet. This is done in the two following ways.

External fibrillation increases the surface area and sheet strength by easing the fibrils from the surface of the fibre, to increase cohesion and contact area between fibre surfaces, leading to improved smoothness, porosity, and tear [1].

Internal fibrillation collapses the fibres and allows water to penetrate and make the fibres more pliant. The flatting increases the sheet strength to create a smooth surface that is less porous. Examples of papers that have not undergone internal fibrillation are blotting paper and hand-made papers, whereas fibres that have this treatment include tracing paper or greaseproof paper. Fibrillation is likely to have a more significant impact in mechanical than in chemical pulping methods [2].

2.1.2.3.6 Recycling paper fibres

As a primary objective in a circular and sustainable economy, used paper and card is recycled as much as possible into secondary fibres, reducing the need for virgin fibre for low-cost products such as packaging and cartons, and increasing sustainability. However, the process of returning waste stock requires extra processing, energy, and additives.

Deinking. As part of the recycling process of paper such as newspapers, magazines or photocopy paper, different chemicals are used to remove ink from printed paper. Enzymatic deinking uses sodium hydroxide to break down ink by hydrolysis to aide ink dispersion. Flotation deinking involves the addition of surfactants and air. Small bubbles release the ink particulates to the surface in a froth which is removed. Bleaching agents are used to increase the brightness of the pulp.

Detrashing. To disperse secondary fibres such as recycled paper, detrashing using a ragger is required to remove unwanted debris, including grit, plastic, wire, glue and metal before entering the flow line and causing damage to equipment.

As paper is recycled repeatedly the fibre strength is weakened. Secondary pulp or deinked/recycled pulp tend to be used for low value moulded products such as drink trays, cup carriers, wine shippers, egg cartons, egg trays, and fruit trays. To improve their strength and water resistance, moulded pulp products can be a sprayed or dip coated with wax [3].

2.1.2.4 The paper-making process

Since the late eighteenth century, industrial paper machinery has evolved constantly, with increasing speed and standards of quality. From the bales of dry pulp sheets, the next step is the creation of a sheet a paper. The aim of the paper-making process is to convert batches of material ratios (paper, pigment, water, chemical additives) from a wet paper pulp into a homogeneous dry continuous sheet of paper. All these steps are undertaken as a continuous process that controls the flow of fibre of a single paper machine. The most used today are commonly called Fourdrinier machines (patent 1806). Of primary importance is the creation of an even distribution of fibre that has no contaminants, such as bugs, air holes or flocs. The removal of water takes up the most energy and therefore any steps that can be taken to use natural drainage and removal of water makes the process more energy efficient. Pulp entering the wet end of the machine usually consists of around 99% water, and at each section the water is efficiently reduced, so that by dry end of the process the water content is under 1%.

2.1.2.5 Stock preparation

Stock may be delivered either as bales of dry pulp from external sources, which are then re-pulped with water to create a slurry, or if the paper has been produced on-site, thick stock is discharged from a pulp storage container and diluted to create the slurry. The dilution of water to paper varies from 0.3% to 3% depending on the weight of paper or board being manufactured. The thicker the stock, the heavier the paper. Weight varies from 12 gsm for a tissue paper, 120 gsm for office photocopy paper, 180 gsm for writing paper, 250 gsm for art paper, or 600 gsm for paperboard.

2.1.2.6 Dispersing the fibres

For dispersing virgin fibre stock that has arrived from the paper mill, low consistency pulpers containing low profile rotors, which are high sheer devices, are good for breaking up wet-strength papers, but can damage the fibres. A high consistency pulper, which has a high profile rotor, is a low shear device, and is not suitable for processing products that have a wet-strength. The slurry then passes through a machine that cuts and de-clusters the fibres to improve fibre bonding properties.

The fibre is consistently controlled and refined, and where any contamination is collected, and air bubbles are removed. Contamination is a significant factor in the overall quality of the finished paper. The fibres are cleaned using screens, by heating, ánd centrifugally. Air pockets and bubbles are important contamination factors. Bubbles can stick to fibres causing fibres to floc and stick to each other, thus reducing tensile strength, or bubbles may burst resulting in a pin hole in the paper.

Pressure screen cylinders and panels are used to remove contaminates that are the same size but of different density. Centrifugal action is used to remove large items where heavy particulates are pushed to the outside and fall to the bottom. Air is removed by applying a vacuum under 40 °C–50 °C and boiling the fibres.

2.1.2.7 *The paper-making process*
There are three components to paper-making: the stock preparation, the wet end, and the dry end. The removal of water takes up the most energy and therefore any steps that can be taken to use natural drainage and removal of water make the process more energy efficient. Pulp entering the wet end of the machine usually consists of around 99% water and at each section the water is so efficiently reduced that, by the dry end of the process, the water content is under 1%. Figure 2.17 shows a simplified version of the paper-making process from start to finish,

2.1.2.8 *Stock preparation*
In large machine chests, fibre is mixed with fillers (coloured pigments, clay, calcium carbonate and titanium dioxide) to form a stock slurry of fibre pulp. The pulp is kept under constant turbulence to maintain consistency and reduce flocking. These chests maintain stock for 15 min retention, which gives enough time to allow for any modification to the consistency or colour of the paper. Within this system, the machine chest stock is referred to as 'thick stock' and the diluted stock is referred to as 'thin stock'. Thick stock describes a consistency greater than 1% and thin stock describes dilution to >1% fibre content.

The approach system extends from the machine chest to the headbox. Forming is crucial for achieving an even spread of fibre across the sheet and has a strong influence on the final appearance and properties of the paper, e.g. an even distribution of fibre, no flocs, no wrinkles or pinholes, directional tensile strength.

2.1.2.9 *The wet end*
The wet end of the machine comprises a flow-spreader (also described as the distributor or header) which takes the pipeline stock to the headbox, and the headbox then jets the fibre evenly across the width of the machine onto a moving web. In many modern paper-making machines, the flow-spreader is an integral component of the headbox, which controls the turbulence, maintains the angle of the fibre, and ensures the stock is delivered consistent with the speed of the web. As the wet web passes through a series of rollers and presses, additional water is removed and the web is consolidated into a uniform and continuous length of wet paper.

2.1.2.10 *Forming*
At the wet end, the thin stock is pumped into the headbox (or flow box, stuff box), whilst maintaining a constant movement and pressure to reduce clumping or

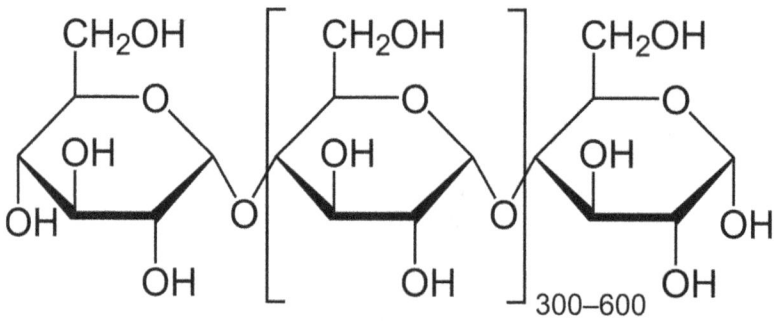

Amylose

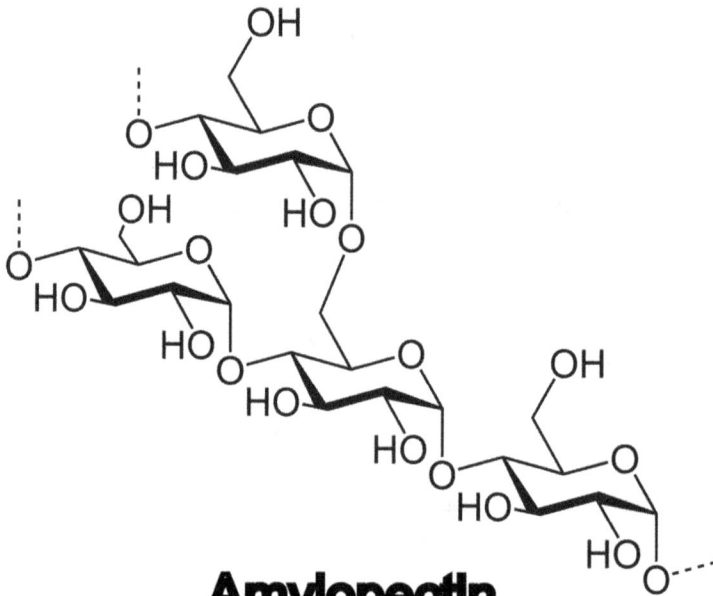

Amylopectin

Figure 2.17. Structure of amylose and amylopectin. (Adapted from [145]. CC BY 4.0.)

flocculation. The function of the flow box is to convert the flow from the cross direction to the machine direction.

Flow boxes have evolved over time—from an open flow box, to pressurized, hydraulic (with air gaps)—to the now more common method of a dilution process. A series of valves jet water to regulate the weight and dilution of paper, thus ensuring an even profile across the whole width of the paper. The thickness and velocity of the deposition needs to be uniform across the full machine width as this determines the final thickness and weight of the paper.

The stock is jetted through an adjustable slice under uniform pressure onto the moving continuous mesh loop. The consistency of the fibres is controlled by the slice opening. The slice comprises a top lip and an apron (bottom lip). The apron is angled slightly and can be adjusted in relation to the speed of the wire.

The jet of stock is deposited at a similar speed to the movement of the wire. The efflux ratio determines the speed of the stock coming onto the wire compared to the speed of the wire and will affect the rush and drag of the fibres, and which, importantly, determines how the fibres are orientated. For example, if the jet of thin stock is deposited faster than the speed of the wire, the fibres tumble over themselves creating a more random orientation of the fibre, which is known as *rush*. This orientation leads to a tensile strength in both directions and is termed as a square sheet. In contrast if the stock is jetted from the slice slower than the wire, fibres are sped-up onto the wire and fibres are *dragged* along. These fibres are orientated in the direction of the machine, which provides a good machine direction tensile strength but poor cross directional tensile strength.

A simple method to determine how a paper has been made, based on the tensile strength and orientation of the fibres, is to tear a piece of paper. Newsprint and toilet paper, for example, has an exaggerated directional tensile strength in one direction (machine direction), whereas greaseproof paper or photocopy paper demonstrates tensile strength in both directions.

The wire or forming fabric is a finely woven mesh screen. It was traditionally made from phosphor bronze, but is now made of a synthetic fabric, polyester monofilaments, which can last up to ten times longer. For the purpose of this explanation it will now be referred to the *wire*. A coarse wire enables faster drainage but creates a coarser paper. The wire has three functions:

1. To transport wet fibre.
2. To permit drainage of water.
3. To transmit power.

The web is the continuous mat of fibres that is in the process of forming or which has already formed the final paper.

2.1.2.11 Formers
There are four main classes of former: cylinder vat formers, Fourdriniers, top formers, and gap formers.

2.1.2.12 The Fourdrinier
Fourdrinier formers (see figure 2.18) were dominant from 1800 to the 1960s. As the stock comes out of the flow box, a jet of stock is deposited onto the wire to form a wet stock of fibrous substrate. The wet substrate then moves across a series of rollers and formers to de-water the wet substrate as energy efficiently as possible. At the beginning of the section, dewatering occurs without a vacuum using rollers and foils, and towards the end of the section dewatering is assisted by suction and vacuum. The drainage from the wet stock is known as white water, regardless of colour, and is then recycled.

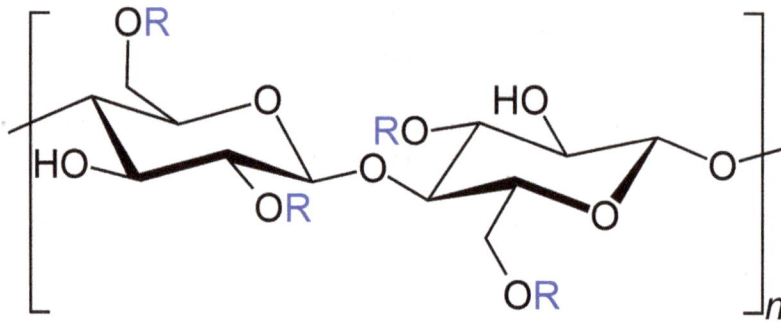

Figure 2.18. Structural formula of carboxymethyl cellulose. (This image has been obtained by the author from the Wikimedia website where it was made available by Jü (2015). It is stated to be in the public domain. It is included within this chapter on that basis. It is attributed to Jü.)

The first section, called the drainage section, comprises two large rolls at each end of the wire loop. The first roll, known as the breast roll, supports the wire, and as the wire comes into contact and curves around the breast roll, it creates sufficient tension to provide substantial dewatering pressure.

As the web moves from the breast roll to the couch roll, the web moves over a series of small diameter table rolls that create an outgoing nip and vacuum. The wire then passes over a series of fixed contoured foils which are sloped blades that skim the under surface of the wire, and by generating small pulses form a hydrodynamic vacuum and achieve drainage of white water through the wire.

These small pulses also assist in further orientation of the fibre. As the web continues to the far end of the wire, white water continues to be drained over a succession of low vacuum boxes and suction boxes. The roll at the far end of the wire, known as the couch roll, is the point where the wire is looped back to the breast roll, and the fibrous substrate is moved on to the next section. By the end of this section, the water content is typically about 60%–70%

At the couch roll, the fibrous substrate is transferred by a pickup roll to the press fabric or felt and onto the press section. During the wire's return journey to the breast roll, there are a series of stretch rollers to control the tension, and a guide roll to detect and adjust the path of the wire to ensure the wire does not move off the rollers, and water jets to ensure the mesh remains clean of debris.

A relatively small number of machines include a dandy roll, which is the only roll that sits above the wire. This roll has a wire covering and a raised pattern stitched on to the wire, which is used to apply a watermark on to the paper. Typical uses of watermarks include banknotes, security documents, high-quality writing paper, and fine art paper.

An issue with the Fourdrinier is that drainage occurs only on one side. Modifications to improve drainage of the water include:

- An inclined wire, to assist in the removal of water from heavy pulp.
- Using twin wire, which is useful for a two-sided sheet, where water is sucked upwards and downwards, increasing drainage and paper production.

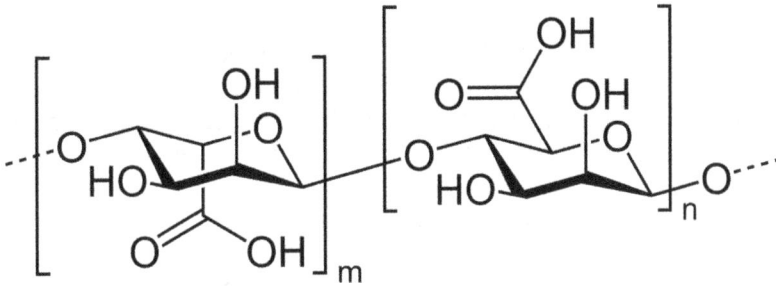

Figure 2.19. Structural formula of aliginate. (This image has been obtained by the author from the Wikimedia website where it was made available by NEUROtiker (2008). It is stated to be in the public domain. It is included within this chapter on that basis. It is attributed to NEUROtiker.)

In the twin wire Fourdrinier, which is useful for lower grammage sheets and in the creation of two separate sheets or a two-sided sheet, white water is jetted from two flow boxes. In Veriformers, as used by the newspaper industry, stock is jetted upwards and the fibres are caught between two wires.

Another former is the cylinder mould machine, invented by John Dickinson in 1809 [4]. Cylinder vat formers or mould machines (figure 2.19) are less common in paper-making and are mainly used to make heavy-weight artists paper for printmaking, painting and watercolour using different high-quality fibre such as cotton.

The former has a large hollow wire-covered cylinder of 260 cm in circumference and 130 cm in width, partially immersed in a deep vat of diluted stock, which slowly revolves. As the cylinder rotates out of the vat, the fibre is collected onto the wire to form a wet web. The constant turbulence of the slurry in the vat ensures uniform density and thickness of the sheet. As the cylinder rotates upwards, the water drops through the wire and is carried away by a vacuum from the inside of the rotating cylinder.

The size of the cylinder is the chief factor in determining the size of the mould-made paper. The speed of this machine is much slower than that of the fast moving Fourdrinier, of between 4 and 15 m min^{-1}.

Resembling the characteristics of traditional hand-made paper, paper fibres fall onto the web in a random orientation, improving the strength and consistency. The paper made on a cylinder mould paper machine presents different characteristics to a Fourdrinier machine, which includes heavier grammage from 80 and 600 g m^{-2}, high caliper or thickness, more stiffness in the machine direction, a matted three-dimensional surface appearance, and irregular deckled edges. Watermarks and metal dividers are also attached to the surface of the cylinder mould with copper wire to mark the size of individual sheets such that, when dried, the sheets can be torn along the marked, weaker lines.

As the cylinder reaches the highest point of the rotation, a horizontal felt is pressed against the top side of the cylinder with a press roll, which further de-waters the web. The web is pulled away from the cylinder and moves to a wire mesh conveyor comprising additional rollers that dry and calender the paper.

2.1.2.13 The pressing section

The objective of the second part of the paper-making process—the pressing section— is to remove water at the same time as consolidating the fibres of the fibrous web to form a sheet of paper. It is important to ensure the fibres are in close contact to form a strong fibrous bonded substrate. Pressing has an important influence on the physical properties as well as the appearance of the paper. As the goal is still to achieve the most economic method possible of dewatering, a series of roll press nips with increasing pressure are used. The wet sheet is compressed in each press nip at linear loads ranging from 30 to 200 kN m^{-1}.

The point at which the press fabric or felt and paper come into contact with the nipping roller (figure 2.20) is where the greatest amount of water is removed. There are four stages in the nipping process. During the first stage, the sheet begins to enter the roller, air begins to be forced out and the sheet is saturated with water, the pressure from the rollers begins to force out the water from the paper to the felt, which then becomes saturated in stage two. By the end of this mid-nip phase, there is a maximum pressure on the paper, and in stage three the nip expands until the hydraulic pressure in the paper is zero, corresponding to the point of maximum paper dryness. The paper and felt leaving the nip both become unsaturated in stage four [5].

In order to prevent water returning back into the paper, an important consideration is to effectively provide the shortest path to follow for water to escape from the nip. Different methods have been explored to achieve the most efficient water removal (figure 2.21). These include a suction method to one side of the paper, or a double felt thus removing and absorbing water on both sides. Other approaches include the use of grooved rollers that can quickly channel away the water from the felt, or drilled holes and suction box inside the roller. Another approach has been to replace one of the rollers with a hydraulically loaded shoe. The shoe is concaved which matches the contour of the press roll, the contact with the felt and paper is 10 times more than conventional press rollers.

Pressing may also include finishing and smoothing the paper. For example, hot pressing a sheet increases dewatering as well as sheet consolidation, and heat also softens the fibres and aids compression. Steam showers are applied to one side of the sheet whilst a vacuum is applied to the felt side. Steam is sucked through the fibres. Sheet pressing is conducted at around 60 °C–90 °C.

As the substrate passes from the forming section to the pressing section it is squeezed though the nips, and then passes on to the drying section.

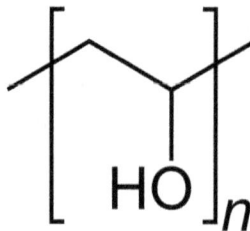

Figure 2.20. Structural formula of polyvinyl alcohol. (This image has been obtained by the author from the Wikimedia website where it was made available by Jü (2015). It is stated to be in the public domain. It is included within this chapter on that basis. It is attributed to Jü.)

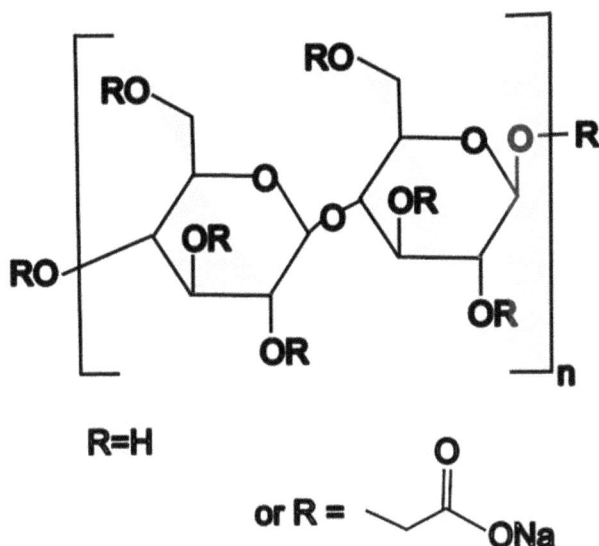

Figure 2.21. Structural formula of sodium carboxymethyl cellulose.

2.1.2.14 The drying section

The third section comprises the dry end, where the remaining water is evaporated through a series of steam-heated cylinders, which dry and press the fibres to create fibre-bonds. The thickness and surface of the sheet is calendered through a series of roll nips. The very large dryer section is the most energy consuming and expensive in terms of capital cost. Therefore, much effort is made to conserve energy and increase evaporation rate by reducing steam usage and the number of dryers. The drying sections are usually enclosed to conserve heat. The evaporation drying rate is measured as weight of water evaporated per hour per area of surface dryer ($kg\ h^{-2}\ m^{-2}$). A balance is required between a high evaporation rate and equipment requirements, but also based on the characteristics and constraints of the paper. For example, bleached grades dry more easily than unbleached and recycled fibre dries more easily than virgin fibre [7].

There are different types of dryer sections, which depends on the paper machine and the desired quality of paper. Multicylinder drying is generally used for printing paper and board grades.

Yankee drying is used for tissue grades and machine-glazed (MG) papers. Yankee cylinders are traditionally made of cast iron and have diameters up to 6 m. Doctor blades can be added to make crepe paper. Through air drying (TAD) is used for tissue grades which avoids compacting the material and is used for bathroom tissue and paper towels.

Different processes come with their own set of pros and cons, for example, the elimination of the condensate that forms inside the large Yankee cylinder, and the TAD process represents a higher energy consumption compared to wet pressed.

The principles of water removal are the same in all set ups. After the wet web has left the press section, it contains about 50%–60% water [8]. The partially dewatered sheet then enters the dryer section, which consists of a series of large steam-heated cylinders or dryer cans (between 150 and 180 cm in diameter) arranged in groups. These cylinders rotate, and the paper is threaded around them in a serpentine path. As the paper travels over the hot surfaces of the dryer cans, the heat evaporates the remaining water in the paper. The cylinders are heated indirectly via steam-filled pipes. The temperature is carefully controlled to prevent scorching or damaging the paper. An air layer between the sheet and cylinder creates thermal resistance, i.e. the air insulates the paper from the heat of the cylinder. The paper must therefore be kept under tension to maximize contact between the dryer can and the paper surface. Evaporated water is driven away from the paper surface by dryer hoods or blowers to prevent condensation and re-wetting.

Drying by heated cylinders can cause problems for coated papers where contact with the cylinder can cause the paper to stick. In that case air-impingement drying or infra-red drying are used [8]. In air-impingement drying high-velocity air jets are directed onto the surface of the paper. When the air hits the surface, it heats it up and causes turbulences which remove the resulting vapour. Air-impingement drying allows rapid drying, precise control of temperature, is energy efficient and provides excellent control over moisture distribution across the paper width, resulting in uniform drying and high-quality paper products. Infra-red drying means that infra-red lamps are positioned across the whole width of the paper web. The infra-red radiation is absorbed by the water and the fibres of the paper and raises the temperature above the boiling point of water, which causes the moisture to evaporate. The vapour must be removed by a ventilation system. Infra-red drying is contactless, energy efficient, fast, uniform and can be easily switched on and off.

Immediately after the drying section, surface sizing can be carried out using a starch solution which can be pigmented if required. This prevents fibres shedding from uncoated surfaces and improves surface strength and smoothness.

Calendering, which is a type of ironing process giving a uniform thickness and smoothness to the paper or paperboard, takes place once the substrate has been sized. A smoother sheet improves printing and uniform thickness is advantageous for the winding process.

The dry substrate is passed between cylinders, which can be cold or heated and water may be applied to enhance the smoothing effect. The cylinders on paper machines are often a combination of steel rolls and ones made of composite material to provide very smooth glossy finishes. The high pressure in the nip between two calender rolls flattens higher parts by deforming the wood fibres on the surface of the sheet permanently. Fibres inside the sheet are also deformed which results in a thinner sheet and a more compact paper [9].

Some papers are produced by 'super calendering' (e.g. glassine papers), which is carried out on a separate machine having up to 14 rolls, to produce a translucent paper or papers which are suitable for magazine printing and cheaper than coated versions. The dry, calendered sheet is wound onto a reel.

Where required, the web is now coated with a white mineral pigment base (clay or chalk). This gives a hard smooth surface suitable for high-quality printing. There are many ways to add the coating and, depending on the colour of the web (brown, grey or white), between one and three separate coats are applied. The amount of coating is also governed by the final smoothness required and the initial smoothness of the web. The smoother and whiter the web initially, the less coating is required. Depending on the end product, the paper is additionally embossed to produce a decorative surface, or cut into sheets.

2.2 Printing on ceramics and metal, from tableware to tiles

Printing colours onto ceramic and metal materials requires a permanent fusion between the material and substrate. The primary reason for using these materials is that they provide tough containers that are much more robust than fabric and paper, are impervious to spoiling and pollutants, and can be cleaned and sanitized, and the contents inside glass or ceramic containers can thus be stored and protected. The products and applications using these materials are numerous, ranging from technological to surgical, sanitary to decorative.

Colour pigment is applied as a transparent glaze coating, an all-over slip coating, direct print, or a transfer decal. A glaze is chemically formulated to adhere to the clay or ground. The three essential elements of a glaze are silica, the glass-forming elements, and flux. The final colour appearance depends on the minerals and oxides, elements and compounds, the translucency, the number of layers of glaze, the firing process, and the melting point of the elements. Glazes may be coloured by oxides that are dissolved into the glaze. These glazes include oxides of iron, copper, manganese and cobalt.

2.2.1 History of the transfer print

For centuries the primary method of creating repetitive marks to create a pattern for decoration was the use of wooden natural or hand-made stamps to impress into the surface of the clay. Primitive woodcuts on paper were applied simply by pressing onto the clay. By the fifteenth century, metal engraved plates were made in roller presses. Most colours would have been carbon black or oxides to increase the range of colours. Cobalt blue was first seen on Chinese ware and became popular in Dutch or English 'Delftware' followed by purple, yellow, green and orange.

2.2.2 Printing onto ceramic

Throughout history, many materials such as glass and ceramic and metal have been used to preserve, contain and transport products, and with the advent of the industrial revolution and the development in methods of mechanization, mass production, and transportation, more goods and products could be manufactured, that also needed to be transported around the country. Food has been one of the most important items in need of storage, containment, and transportation. Early containers were large earthenware jars covered with a cloth, however the size of the jars and exposure to air meant the food inside quickly became mouldy. As is self-evident to contemporary consumers, the labelling of containers was vital to inform

Figure 2.22. A modern version of a tissue transfer print: *Reinstated shard: quiet work*, 2022, reproduced by kind permission of Lisa Sheppy, 2024.

customers of what was inside. Paper labels pasted onto containers could quickly become detached, so alternative methods of permanent fixing were developed by fusing labels onto the pots using heat.

The origin of paper-based transfers for ceramic was arguably in the mid-eighteenth century, either in the Doccia Factory (John Copeland, Spode's Willow Pattern), or in John Sadler of Liverpool in 1749. The process was called 'bat printing' .This involved printing from a thin piece of tissue paper, applied to an unglazed, fired pot or jar, then glazed and fired to fuse the layers to create a glassy and impervious surface (figure 2.22). The black printed underglaze transfer could be printed cheaply, transported in smaller pots and in vast numbers. Small pots for potted meet, anchovy paste, toothpaste, butter, marmalade, oil, etc, were widespread from the 1830s until the 1880s. The bodies of the pots were invariable white, and the lids of the pots were decorated with the supplier's name and the contents. Some pot lids included logos, fancy lettering and decorative edges, allowing the crafts-person to explore design in the round.

Intaglio underglaze tissue is a traditional method that uses etching or intaglio plates, the plate is inked and wiped and printed onto a tissue transfer paper. The inked paper is then pressed onto an unglazed ceramic or biscuit-fired ware (biscuit firing is the first firing of the raw clay, firing occurs at 900 °C–1000 °C, the biscuit body is porous and absorbs glaze and colour).

The type of tissue paper used, sometimes referred to as silk paper or silver paper, was developed in the 1830s by Henry and Sealy Fourdrinier to produce a highly robust product for the pottery industry that could withstand the rigors of printing and wetting without tearing. The traditional pigment cobalt oxide, when it reacted with the glaze, resulted in the traditional blue colour ware.

The printing equipment and the process in the nineteenth century involved a hot iron surface upon which the copper plate was heated to keep the ink workable and an etching press to impress and transfer the ink to the paper (see also intaglio printing). A thin sheet of tissue paper, dampened with soap and water, was placed onto the etching plate. The purpose of the potash soap made from vegetable oils and water was to prevent the tissue from being scorched by the hot copper plate and to enable the paper to remain pliable when applied to the surface of the pot. The inky lines and marks were impressed onto to the tissue using a press. The plate was returned to the hot plate to help release the inked tissue and prevent the paper from tearing. The printed tacky tissue was kept facing upwards and surplus paper was trimmed to the fit the dimensions of the pot. The tissue was applied tacky ink down onto the yet unglazed, absorbent biscuit-fired pot, and the back of the paper rubber vigorously with a stiff brush to thoroughly transfer all the ink to the porous surface.

The paper was then dampened and peeled away. The pots were moved to a hardening kiln that burnt away the oils, achieved by firing at 680 °C–750 °C equivalent to 'red heat', which was important to burn out the medium, for the metallic oxide to remain, and fuse the ink to the biscuit-ware before the glaze was applied. Glaze was then applied by dipping the ware into a vat of liquid glaze— finely ground glass suspended in water—and re-fired to obtain a glassy hard surface with an image fused between the layers.

The use of paper transfers led to the 'Staffordshire blue' pottery or blue ware. Cobalt was the main colour, and many different blues were created. The term for the blue ware describing the cobalt colour also varies, and is variably described as Mandarin blue, garter blue, Arabian blue, smalt, ultramarine, royal blue, cyanine, Delft blue, Ming blue, and Nankin blue.

Thousands of different patterns were created. A number of different styles, including the Willow pattern, are still recognizable today.

In the 1950s an industrial development called the Murray Curvex system enabled greater mechanization for printing decoration onto flatware such as plates and saucers [40]. It is a flexible gelatine pad that is inked from a plate and then pressed onto the surface conforming to the shape of the plate. Pad printing, as it is called today, is still in use for printing onto curved or irregular surface. The gelatine has now been replaced by silicone (see pad printing).

Figure 2.23. A modern example of the Willow pattern, 'Willow Blue', made by Johnson Bros, England.

Contemporary methods of printing by the transfer process have changed little and require a great amount of hands-on work, as exemplified by the Victorian heritage company Burgess and Leigh in Stoke-on-Trent, who still employ this method.

One of the most enduring patterns that has retained its popularity since the nineteenth century is the so-called Blue Willow pattern ware. Its allure has been the distinctive cobalt blue colour and the iconic elements that are recognizable in the pattern, these include the figures crossing the bridge, flying birds, a tea house, stylized trees including a willow tree, a fence and a garden, a small boat and a sailor, and decorative oriental style edging in line work (see figure 2.23).

The original image was believed to have been made by Thomas Minton for Thomas Turner's Caughley factory in Shropshire in 1780, and was called Willow Nankin. Minton moved to the Spode factory, Staffordshire, in 1784 and Josiah Spode developed the design based on a Chinese pattern called Mandarin. Spode's version may have been produced around 1790, the engraving was all by line, and later versions included stipple and punchwork—a method of hammering dots or shapes into the metal to create patterns—suitable for more organic areas such as the trees and garden. The production of Willow ware evolved very quickly to encompass a wide range of different patterns onto porcelain, pearlware, bone china, and robust earthenware and stoneware.

The ceramics industry in Britain evolved quickly and the transfer ware process enabled complex patterns on white or cream earthen or stoneware to be made cheaply. By the late 1800s many ceramics companies produced many variations of the Willow Pattern.

2.2.3 Printing of enamel

Vitreous enamel is a process of chemically bonding vitreous materials to metal that are melted or fused by heat to the surface and to ensure a robust and glazed surface. The process of bonding molten glass to metal is critical to ensure colours, patterns and images are fixed and will not fade with exposure to light or weather. The colours are typically applied by screen printing, spraying, rollers, or water-slide decals that are laserprinted or screen printed.

The Italian word 'frit' describes a glaze ingredient that has already been melted to form a glass, crushed into powder, suspended in water and applied as a decorative or protective layer to the surface of an object. It is scratch resistant, durable and colourful.

Enamel has been used to fuse to metal for more than 4000 years. The process of fusing coloured powdered glass to metal has been used to create brilliant jewel-like artefacts including crowns and religious reliquaries, fabulous Fabergé eggs, watch faces, and delicate items of jewellery. On a day-to-day level, enamelled cups, plates and stoves give a robust and sanitary product that is resistant to breaking and chemical erosion.

With the introduction of the railway in the 1830s and the London Underground in the 1860s, enamel signs began to be produced in their thousands. Developments in printing methods meant that new styles and forms of signage (figure 2.24) and

Figure 2.24. Enamelled advertising in the Museum in the Park, Stroud.

lettering evolved, and some of the classic brands can still be identified today. In 1889 the Patent Enamel Company set up a factory to create tough, coloured signs suitable for all weathers. Different methods were used including stencilling, rubber stamping and lithographic transfers. In the 1920s a range of signs were produced using the new silkscreen method and printed onto steel, which made them easier and quicker to produce with brighter colours. By the 1950s steel shortages marked the demise of the enamel sign.

Enamel is applied to a metal base and fired until the glass is molten. Enamelled surfaces typically contain 10%–15% coloured pigments by weight. The firing time varies from 2–10 min and the temperature varies between 760 °C and 900 °C depending on the surface area and the desired finish. The melting point is dependent on whether a hard, medium or soft enamel is required, or whether a pre-enamelled or industrially enamelled surface has been used, eg. enamelling decal numbers onto a cooker where a hard firing enamel may need to be used.

The aim is to ensure uniform melting and to adhere to the surface without damaging the object, which may involve different firing methods. For example, to build up layers of colour on metal, the coated object is placed in a preheated furnace and fired from 1 to 5 min (depending on the size or technique). Once the object is removed and allowed to cool to room temperature, further coats or colours are applied and many firings may be required to create the desired finish. For fusing enamels to glass, the object is placed into a room temperature kiln and the heat is steadily increased, then reduced to an annealing temperature and maintained for a period, and then further reduced to room temperature before opening the kiln to prevent stresses and shattering.

Transfer decals should always be applied to a pre-enamelled ground. The process of burning off is essential to ensure there is a bond between the decal and the pre-enamelled ground. The item with the decal is placed in a cold kiln and once the temperature reaches 400 °C–450 °C the item is removed and placed in a second kiln preheated to 760 °C. As the decal layer is very thin, the fusing process requires a brief firing.

Contemporary uses of enamel are numerous, and the tradition has continued well into the twenty-first century. Companies such as AJ Wells in the Isle of Wight continue to print underground signs for Transport for London, produce highly coloured metal cladding for architecture and waymarking signs for many cities from Bristol to New York. As well as being resistant to weather, these panels are also graffiti proof, temperature resistant, and fully recyclable.

The tighter restrictions on health and safety and environmental protection have required that most pigments are lead free (with the exception of some sphenes), and in place of some pigments such as cadmiums other materials using zirconium silicate have been used to obtain the more saturated reds and oranges.

Heat resistant pigments for enamel are listed in table 2.6.

Table 2.6. Pigments for enamel.

Material	Formula	Molecular weight	Colour appearance and notes
Antimony	SB_2O_3	288	Antimony is used to produce the colour Naples yellow. In the glass industry small percentages of antimony oxide are used to remove bubbles in optical glass, to decolorize specialty glasses, and as a stabilizing agent in the production of emerald green glass.
Chromium oxide	Cr_2O_3	152	Chromium oxide is a versatile colouring oxide that ranges across a range of hues including red, yellow, pink, brown, and green.
Cobalt oxide	Co_2O_3	241	Cobalt oxide is one of the more stable colourants and gives a similar shade of blue across firing conditions, thus enabling a consistency of appearance.
Cobalt carbonate	$CoCO_3$	119	A red-pink powder, produces blue in glazes and gives a more evenly distributed colour than cobalt oxide.
Copper oxide	Cu_2O	80	Copper oxide has been used since prehistory to produce glazes that range from blue to green. Some of the earliest examples of pottery were made in Egypt around 3000 BC of the familiar blue faience associated with Egyptian and Persian ceramics. Copper oxide is highly soluble in glazes.
Copper carbonate	$MnCO_3$	115	Light green.
Manganese carbonate	$MnCO_3$		Pink, deep violet—plum colours.
Manganese dioxide	MnO_2	87	An inorganic compound, it occurs naturally as the mineral pyrolusite, and is blackish or brown in colour.
Manganese oxide	Fe_2MnO_4	86	Gives a purple colour, which ranges from a rich blue to plum when mixed in an alkaline glaze. Combined with other oxides gives a cool brown or when mixed with cobalt oxide produces deep violet or plum colours.
Nickel oxide	NiO_2	75	Nickel oxide colourants range from yellow, to green to brown, depending on its content. It can be unpredictable and hard to repeat. The oxide can be combined with other metal oxides to produce special colour pigments. Also used in glass frits for enamel and to develop colours in clear glass.
Iron oxide	Fe_2O_3	160	The most important and widely accessible pigment, as the Earth contains a high level of iron. Iron oxide can produce a wide range of colours from light yellow, rust colours, reddish brown to dark brown and black. The fine particle size of iron oxide can stain the skin. Soap can remove it but it does not dissolve in water.

(Continued)

Table 2.6. (*Continued*)

Material	Formula	Molecular weight	Colour appearance and notes
Red iron oxide			High iron raw materials; alternative names: burnt sienna, crocus martis, Indian red, red ochre, red oxide, Spanish red. Iron is the principal contaminant in most clay materials. A low iron content, for example, is very important in the kaolins used for porcelain
Yellow iron oxide			Used in paints, enamels, concrete, rubber, and paper where permanent yellow is required. It has excellent hiding power, absorbs ultraviolet light, is compatible with a broad range of vehicles, disperses well in aqueous and solvent systems, and does not contain heavy metals.
Tin oxide	SnO_2	151	Tin oxide is a white or off-white powder and is useful for improving the opacity of glazes and produces stunning effects in many coloured glazes.
Gold, silver, platinum			May be used to create pink, red and purple colours. Platinum used for grey. Silver and bismuth are used to create lustre ware.
Zircon		183	Zirconium silicate or zircon are used by the tile, sanitaryware and tableware industries to improve opacity, often characterized as 'toilet bowl white'. Very stable.
Zirconium oxide		123	
Vandium oxide	V_2O_5		Vandium oxide tends to be used as a yellow stain in glazes.
Rutile			The term 'rutile' is generally understood to refer to the brown powder into which these minerals are ground.

2.2.4 Recent issues in the use of cadmium and its legislation

The principal pigments of cadmium are a family of yellow, orange and red cadmium sulphides and sulfoselenides, as well as compounds with other metals. Cadmium yellow, cadmium orange and cadmium red are familiar names in artists' colours. The cadmium pigments are typically brilliantly coloured, with good permanence and tinting power.

Most people would be familiar with cadmium to make rechargeable nickel–cadmium batteries, which have now been replaced by other rechargeable nickel-chemistry cell varieties such as NiMH cells. According to the European Restriction of Hazardous Substances Directive, cadmium is registered as toxic to humans and other animals even in very small amounts, and especially when it is inhaled.

This may occur when working with a powdered material, for example mixing pigments with oil or gum Arabic to make inks, meaning that any art supplies that contain cadmium pigments are being phased out.

However, because the pigments have many desirable qualities, including their brilliance and resistant to fading, artists still want to continue to use them. In March 2016 an exception was made by the European Commission confirming that it will not adopt a REACH restriction on cadmium in artists' paints [41].

As a result of the new restriction, more than 190 m synthetic chemicals are registered globally and a new industrial chemical is created every 1.4 s on average, meaning that healthier, safer products are being sought. The search for a brighter red pigment is underway, and which is the holy grail for the printing and ceramics industry. New synthetic methods are being sought including laboratory grown red crystals that can then be ground to make brilliant and permanent pigments.

2.2.5 Other printing methods for glass, enamel, and ceramic

Today, decorative coloured decals can be made using a range of printing processes, including screen printing and electrostatic transfer, direct digital printing and pad printing.

2.2.6 On-glaze digital decals

Laser printers are fitted with specially modified ceramic toner cartridges that contain finely ground formulations of frits that are deposited on a specially formulated decal paper. The images are generated on a computer, and laser printers can print a wide range of colours. The laser decal transfer is backed onto a paper substrate and, once soaked in water, is released and provides a malleable film that can be moulded on the surface of a ceramic. These decals tend to be applied onto pre-glazed ceramics, which are then fired to fuse with the ceramic body glaze. The on-glaze decal process greatly speeds up the production, enables products to be made in small batches or print on demand. However, because the fired decals have no surface glaze and are fired at lower temperatures, they produce less durable products than underglaze printed items. Also the colours are less durable and after long-term dishwasher washing there is a noticeable fading of the red pigments. This fading is caused by the removal of metal ions from the coloured areas.

2.2.7 Hand decal printing using screen printing

Screen printing using decals (similar to the underglaze tissue transfer for ceramic ware) involves printing finely ground glass coloured power onto a decal paper which is applied to on-glaze (already glazed) ware (see the screen printing process). Ceramic inks can be purchased from commercial suppliers and it is possible to mix these with water-based acrylic screen media. The colour is mixed with a palette knife or a glass muller in a ratio of one part medium to two parts ceramic pigment. The colour layers are screen printed onto a decal paper. The decal paper is a carrier paper that has been precoated with gum Arabic and a clear varnish called a 'covercoat' which, when dry, bonds to the printed image. The decal can then be cut and placed into warm water. The plastic sheet containing the print will slide away from the gummed carrier paper and can be placed onto the ceramic surface, gum side down—image side up. The decal is firmly pressed onto the glazed surface and any air bubbles are removed using fingers or a smooth flat piece of rubber, called a kidney. When fired the plastic coating is burnt away and the printed image fuses with the ceramic surface. On-glaze decals are fused onto glazed surfaces at 750 °C– 850 °C. Because they are not protected by a protective glaze, as in the tissue transfer, they are subject to abrasion and are less robust.

Alternative methods such as the traditional screen printing methods provide opportunities for specialist printing applications, low-volume production, and increase opportunities for applying materials onto different substrates such as glass and metal that are not available in digital printing.

Recent developments have resulted in high-temperature water-slide transfers suitable for in-glaze custom decorating on to different ceramic ware, such as china and tiles. The pigments can be fired at higher temperatures of 1100 °C–1300 °C to ensure improved fusion to the underglaze, thus improving durability and resistance [42].

2.2.8 Direct inkjet printing

Digital printing methods are becoming increasingly applicable for a range of industrial printing. Overcoming the abrasive issues of vitreous materials, significant developments in the ceramic printing industry have meant that digitally inkjetted printed images can be created on a grand scale using flatbed, super-wide format inkjet printing to produce hardwearing ceramic tiles and surfaces, which are used, for example, in hotels and reception areas to create the appearance of high-quality highly polished marble floors and walls, or the appearance and texture of aged and distressed wood panels. Inkjet ceramic decoration printers use piezoelectric inkjet printheads and ceramic powders. The advantages of wideformat inkjet hardware includes the ability to print at a large scale and at high resolution, and multiple patterns reducing any obvious repeat pattern. Colour matching and 'seed-based' pattern creation can also be used to mimic existing patterns and surroundings. These methods also minimize waste and ecological impact. However, inkjetting vitreous materials is still a relatively niche process and is designed for the high-quality and luxury market, requiring a significant investment return. The particulate sizes of the vitreous materials are highly erosive on the printing devices, resulting in wear and tear.

2.2.9 Pad printing

Pad printing, which still occupies a niche but significant sector of the printing market, is an intaglio process that is used to print on to complex, compound angles and textured surfaces. It is commonly used to print onto small objects that are already fabricated, that may be delicate and require decoration or graphical information (numbers, logos), such as on cooker fronts, coins and medals.

2.3 Printing on plastic

2.3.1 A short history of plastic

In the widest sense 'plastic' means any material which can be deformed or moulded under pressure. Such materials have been known since antiquity, for example horn, tortoiseshell, amber, shellac, and natural rubber [43]. Today by 'plastic' we mean materials composed of long chains of carbon and hydrogen atoms, so-called polymers. The use of plastic as a replacement for more valuable materials, such as ivory for example, and the creation of new composite materials, such as the waterproof fabric for MacIntosh coats, started in the nineteenth century. It took until the 1950s that printing was used in combination with plastic substrates.

2.3.1.1 Natural plastics: horn

Already in 1284 the Worshipful Company of Horners of London was recorded, a guild for all those who used horns of animals to produce cutlery, drinking vessels, buttons, combs, boxes, shoe horns, etc. Today the company still exists and since 1943 has been the guild of all those who work with plastics. Horn is a non-synthetic plastic made of keratin, a group of fibrous proteins which are the building blocks for hair, horn, hooves, scales, feathers, claws, and the outer layer of the skin of vertebrae. Like man-made plastics it is based on carbon chains with hydrogen bonds (see figure 2.25). With increasing moisture and temperature, the modulus (a measure for the elasticity of a material) and strength of horn decrease but its breaking strain increases, i.e. it can be deformed without breaking [44]. This allows one to shape horn when it is wet and warm. When cooled down and dried, it will then keep the shape. An example of shaped ivory can be seen in figure 2.26.

2.3.1.2 Shellac

Shellac or lac is an insect secretion by the scale insect *Kerria lacca*. The larva feed of trees, commonly found in India, Myanmar, Thailand and south China, and secrete a resin, which is collected and cleaned. In the late sixteenth century shellac was first imported into Europe and found applications as an adhesive, sealing, insulating, and coating material. Its chemical composition can be found in [46]. It is a thermoplastic

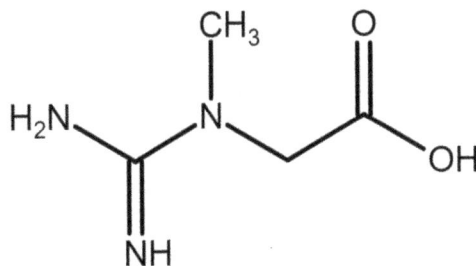

Figure 2.25. Structural formula of neutral keratin. (This image has been obtained by the author from the Wikimedia website where it was made available by Bernardirfan (2023), under a CC BY-SA 4.0 license. It is included within this chapter on that basis. It is attributed to Bernardirfan.)

Figure 2.26. Ivory hairbrush.

with a melting point between 65 °C and 91 °C [47]. Until the advent of vinyl, records were made from shellac [48]. Mixed with wood flour it was used to mould picture frames, jewellery, boxes, inkwells, and so-called union cases which protected daguerreotypes (see figure 2.27).

2.3.1.3 Natural rubber
Natural rubber is indigenous to America. It was mentioned in *La Historia Natural y General de las Indias* by Gonzalo Fernandez de Oviedo y Valdes [49]. *La Historia* chronicles the European expedition to the New World in the years between 1492 and 1549, describing not only the expeditions but also the flora and fauna. The book itself has a complicated history and was only completely published in the nineteenth century [50]. It took more than 200 years until the first samples of rubber were brought to Europe by the French explorer and physicist Charles Marie de Condamine in 1744 [49], who called it 'latex' referring to its milky appearance. It remained an exotic novelty until the early part of the nineteenth century, since unprocessed natural rubber becomes soft in the heat and hard in the cold. The first rubber factory in Europe was established in Paris in 1803 [51], and by the mid -1800s, there were several more factories throughout in Europe. In June 1823 Charles Macintosh patented the lamination process of natural rubber dissolved in naphtha [52] between two layers of fabric. This waterproof fabric is the basis of the Mackintosh raincoat. Thomas Hancock improved the process by mastication [53]. When natural rubber is heated and then worked intensively, it will become more plastic rather than elastic. It can then flow and be shaped. The next step was vulcanization, where natural rubber is heated in the presence of sulphur and other chemicals. Within weeks in 1844 Thomas Hancock and Charles Goodyear [54] were granted a patent for the process. Vulcanized rubber does not become sticky when warm or too hard when cooled. It is still the main material used for tyres. When rubber is heated with substantial quantities of sulphur, vulcanite (ore ebonite in America) is formed. It is the first semi-synthetic plastic. In the nineteenth century rubber was

Figure 2.27. Union case to protect daguerreotypes, shellac covered with leather-imitating paper.

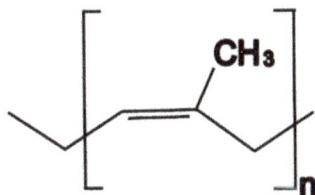

Figure 2.28. Cis-1,4-polyisoprene.

harvested in South and Central America and in Africa. Brazil forbade the export of rubber seeds. Nevertheless they were smuggled to England in 1876 and were the foundation of all rubber plantations in East Asia, to where the centre of world production shifted [51]. Until the 1960s, when it was replaced by thermo-plastics, it was widely used as the material for the boxes of car batteries, combs, jewellery (as a replacement for jet), the mouth pieces of wind instruments, rubber to metal bonding, pens, etc. Today it has become interesting again as a sustainable material.

Natural rubber from the *Hevea* tree varies in composition, but the rubber molecules consist mainly of cis-1,4-polyisoprene (figure 2.28) [49].

Natural rubber is a unsaturated aliphatic hydrocarbon polymer and maybe behave like an alkene with its chemical reactivity dominated by the double bond. During vulcanization the sulphur atoms crosslink the polymer chains.

2.3.1.4 Gutta-percha

Gutta-percha is a close relative to natural rubber. It is obtained from the leaves of the *Palaquium oblongifolium* occurring in Malaysia and Indonesia. Its main component is trans-1,4-polyisoprene (figure 2.29). The latex is more viscous than that of natural rubber, behaves very similar with regard to chemical processes, but deteriorates very quickly when exposed to light and air.

In Victorian times it was popular for decorative items and was used as insulation for undersea cables until the Second World War [49]. Because it does not react with the human body, it is still used in dental root canal treatment [56].

2.3.1.5 Bois Durci

Another non-synthetic material is Bois Durci ('hard wood'), a type of hard plastic patented by Francois Charles Lepage in 1856 [43]. It is made by combining sawdust or wood flour with a binder, such as egg or blood albumen, which is then moulded under high pressure and heat. Sometimes mineral or metallic powders are added too. The resulting material is hard and dense and can be carved, polished, and dyed like real wood. Bois Durci was popular between 1855 and 1890 for use in decorative arts, jewellery, small decorative times such as boxes, frames, figurines and medals. It was prized for its durability and ability to allow intricate detail when carved.

Today, Bois Durci is still produced and used by some artisans and designers.

2.3.1.6 Parkesine

Parkesine is an early form of plastic invented by the British chemist Alexander Parkes in 1856 [57]. It was also marketed as Ivoride and Xylonite. It was the first thermoplastic material to be manufactured commercially and was made by dissolving cellulose in a mixture of nitric and sulphuric acid, adding vegetable oil as a plasticizer, and then treating the resulting material with camphor to make it more pliable [58]. Parkesine could be moulded into various shapes when heated, and when cooled it retained its shape. It was used in a variety of products, including combs, buttons, and jewellery, but was eventually replaced by other types of plastics that were easier and cheaper to manufacture.

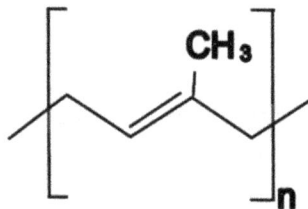

Figure 2.29. Trans-1,4-polyisoprene.

2.3.1.7 Celluloid

Celluloid is a type of plastic that was invented in the mid-nineteenth century and has had a profound impact on the world of photography and film. Celluloid is a mixture of dissolved nitrocellulose or collodion and camphor (see figure 2.30). Its chemistry is not completely understood.

In 1865 the American inventor John Wesley Hyatt improved upon Parkesine by using a mixture of cellulose and camphor and adding heat and pressure to create a more stable material. Hyatt and his brother Isaiah patented his new material and the process in 1870 [59]. In 1871 the brothers registered 'celluloid' as a trade mark [57]. They began manufacturing it on a large scale. Celluloid quickly became popular as a substitute for ivory, which was becoming more expensive and difficult to obtain. It was used to make a variety of products, including billiard balls, guitar picks, and even false teeth. The early years of celluloid production were marked by a number of safety concerns, as the material was highly flammable and was said to be prone to explosive accidents, however, also no major accident has ever been recorded [60].

In the late nineteenth century, celluloid made analogue photography and motion pictures possible. Celluloid was the substrate of choice for Kodak films.

The Eastman Kodak Company was founded by George Eastman in 1888 to sell his roll film camera system. Eastman, in collaboration with William Hall Walker, had invented a new type of dry photographic film and an easy to load camera aimed at the amateur market. In 1888 he offered a camera loaded with a 100-exposure film, then paper based, and an accompanying printing and enlarging service (see figure 2.31). The system was marketed with the famous slogan: 'You press the button—we do the rest'. In July 1888 he showcased the system at the annual photographers' convention in Minneapolis and was awarded 'the invention of the year in photography' medal. The Kodak company quickly became the leading producer of photographic materials in the world. Soon paper as a film substrate was replaced by celluloid [61] based on the patent of Goodwin [62].

The Lumiere brothers, who are credited with inventing the first motion picture camera, used celluloid film to capture their famous short films in 1895 (see for example https://www.youtube.com/watch?v=6Uo3hV-BWv4). Only in 1948 was the highly flammable nitrocellulose film in motion pictures replaced by cellulose triacetate safety film [43]. Until then photographers and filmmakers used cellulose

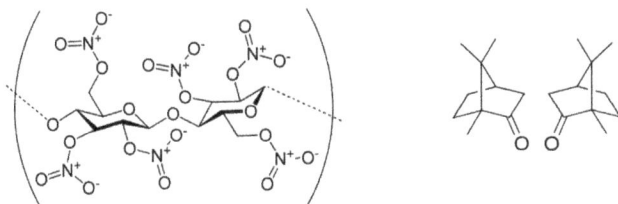

Figure 2.30. Nitrocellulose (left) and camphor (right).

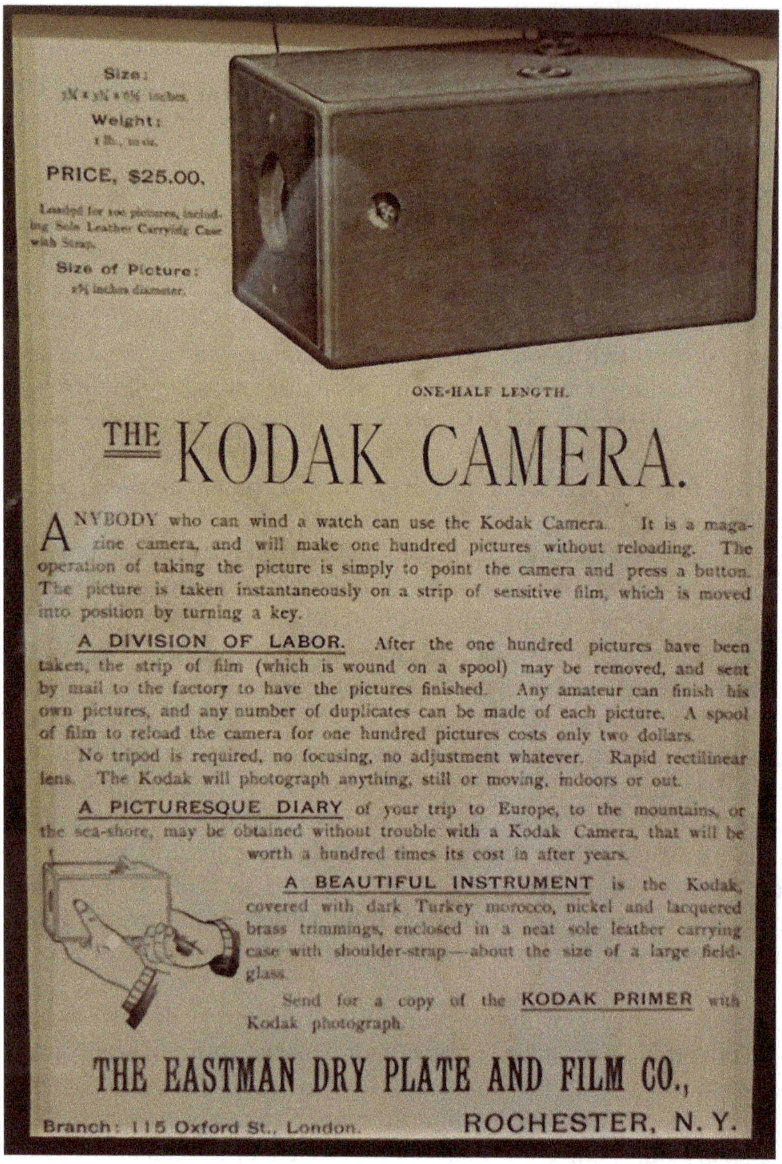

Figure 2.31. Advertising for the Kodak camera with film. (Photograph taken by Susanne Klein at the George Eastman Museum, Rochester, NY.)

nitrate more often than any other substrate. The highly flammable nature of the film led to many fires where the films were stored or screened [63].

2.3.1.8 Cellulose acetate

Cellulose acetate, the non-flammable sister of nitrocellulose, is made by an acetic acid process, when cellulose reacts with acetic anhydride and acetic acid in the presence of sulphuric acid [64] (figure 2.32). First introduced as an alternative to

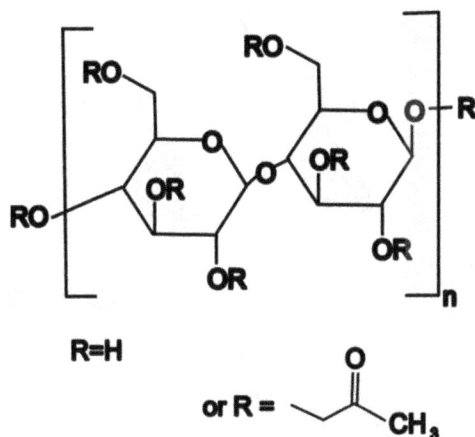

Figure 2.32. Cellulose acetate.

celluloid in the film industry [63], it has many other applications, for example as a material for eyeglass frames, buttons, fabric, filters, films, etc. In museums, the so-called 'vinegar syndrome' poses a substantial risk to collections of photographic materials. When stored at room temperature, cellulose acetate deteriorates releasing acetic acid which then triggers further deterioration in nearby objects.

The deterioration cannot be reversed and can destroy images completely [63].

2.3.1.9 Casein
Casein plastic, also known as milk plastic, is a type of plastic that is made from casein, a protein found in milk. Legend has it that it was discovered by the cat of Adolf Spitteler, a chemist who together with Wilhelm Krische filed a patent about the production of Galalith in 1897. It is said that the cat knocked a bottle with formaldehyde into a bowl with milk which then formed a hornlike substance [65]. The formaldehyde causes the milk protein to crosslink and form a hard, durable material. The resulting plastic is translucent, has a smooth surface, and can be moulded into various shapes. In 1913 1.500 tons of Galalith were produced in Germany using 6% of the total milk production. It was so successful because it was easy to work, easy to dye, could be polished and had a hornlike appearance. In 1914 the production of Erinoid (see figure 2.33), one of the British trade names, started in Stroud, Gloucestershire, mostly for buttons. Casein plastic became a staple of costume jewellery until the 1980 when the production almost completely disappeared because of increasing restrictions on the use of formaldehyde [65].

Although casein plastic was once widely used for items such as buttons, combs, and other small household items, it has largely been replaced by synthetic plastics such as polyethylene and polypropylene. However, casein plastic is still sometimes used in specialized applications, such as in the coating of pills or capsules.

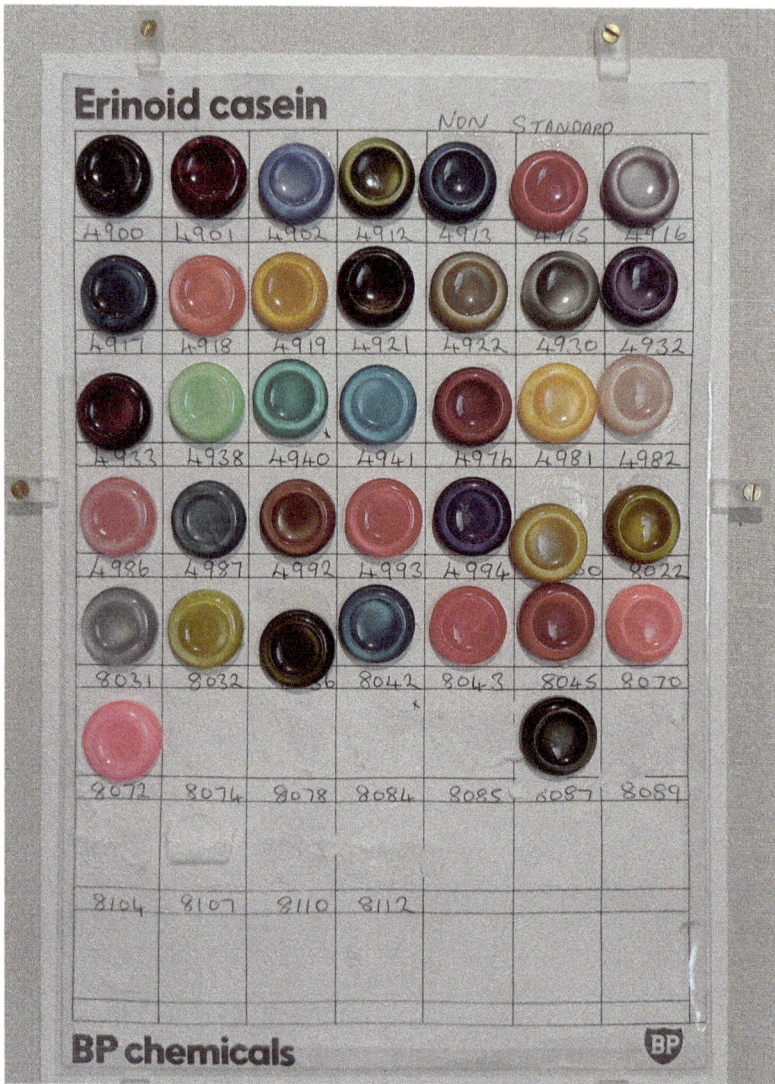

Figure 2.33. Einoid button sampler in the Museum in the Park, Stroud.

2.3.1.10 Synthetic plastics
Synthetic plastics are man-made materials. The process of making synthetic plastics involves polymerization, where monomers, the building units, are chemically bonded together to from long chains or networks.

2.3.1.11 Bakelite
Bakelite was the first truly synthetic plastic (see figure 2.34). It was patented in 1907 by Leo Baekeland [66]. He filed the patent one day before his main competitor, Sir James Swinburne [43]. In the patent he stated that

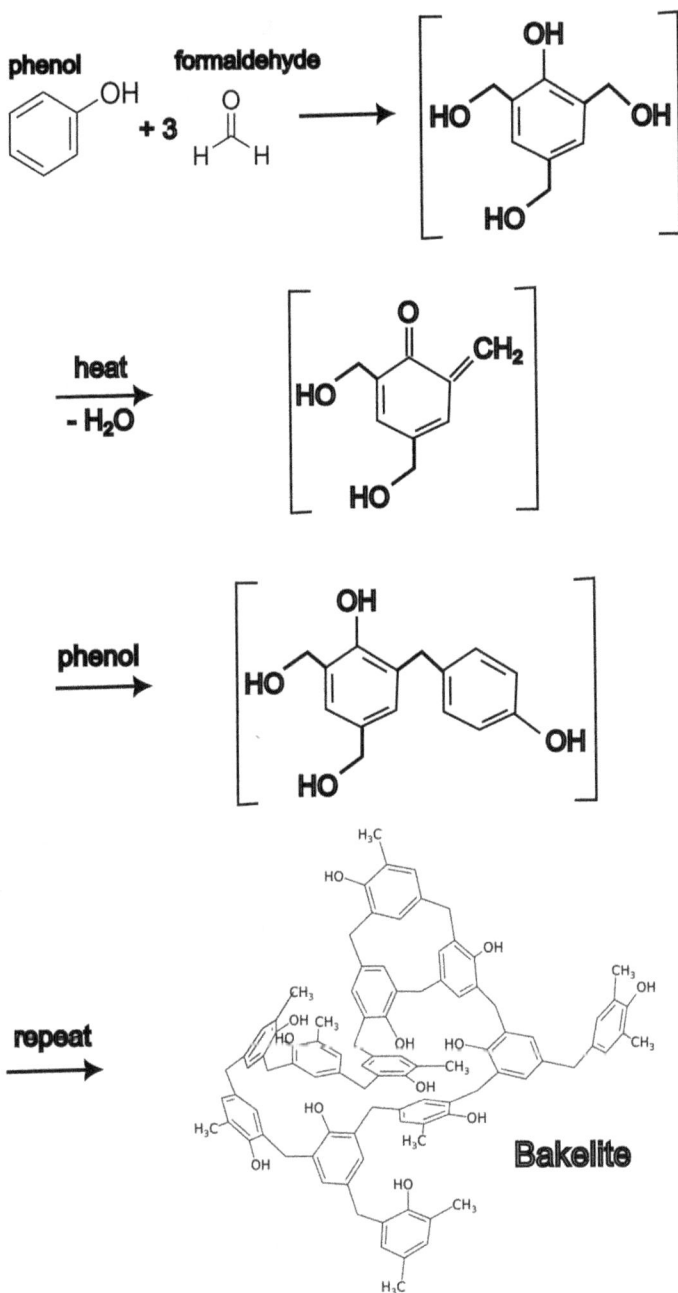

Figure 2.34. Chemical reaction during the production of Bakelite according to [67]. (This image has been obtained by the author from the Wikimedia website where it was made available by MarkusZi (2005), under a CC BY-SA 3.0 license. It is included within this chapter on that basis. It is attributed to MarkusZi.)

If a mixture of phenol or its homologues and formaldehyde or its polymers be heated, alone or in the presence of catalytic or condensing agents, ... approximately equal volumes of commercial phenol or cresylic acid and commercial formaldehyde, these bodes react upon each other and yield a product of two liquids which will separate or stratify on standing. The lighter or supernatant liquid is an aqueous solution,... whereas the heavier liquid is oily or viscous in character and contains the first products of chemical condensation or dehydration. ... If it be desired to mold the material directly the condensation product is poured or pressed into a suitable mold and is submitted there in while maintaining appropriate pressure to a suitable temperature, say about 100–140 °C.; under these conditions there is obtained in from one to two hours or less a hard, compact, perfectly homogeneous mass similar in its properties to hard rubber or to ivory, insoluble in alcohol, acetone, and resistant to heat, ... to moisture and most chemical reagents ... .

From the late 1940s Bakelite started to disappear from consumer products. Its weak point was colour. Bakelite alone is quite brittle. To strengthen it, it is filled with other materials which result in a dull colour [68]. With the advent of different petrol-based plastics, it was replaced more and more by materials with the same features but better colour properties.

2.3.1.12 Polyethylene (PE)

Polyethylene or polyethene is one of the most common types of plastic. In 2018 more than 10 million tons were produced [69]. PE is a group of materials with different properties which are achieved by choosing different initiators and polymerization processes for the starting monomer ethylene, see figure 2.35.

Catalytic polymerization creates a highly linear molecule with high crystallinity and high density—high-density polyethylene (HDPE) (see figure 2.36(a)). Free radical initiated polymerization at high temperature and pressure leads to a high degree of short and long-chain branching which reduces the crystallization in a densely packed material. The result is low-density polyethylene (LDPE) (see

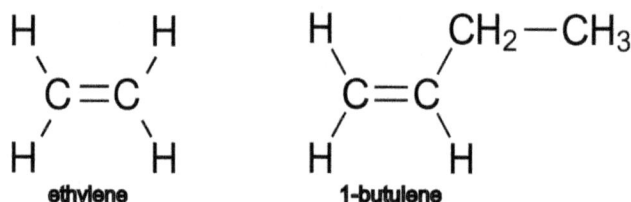

Figure 2.35. A single ethylene and 1-butylene molecule used in the production of LLPE. (This image has been obtained by the author from the Wikimedia website where it was made available by JaGa (2008). It is stated to be in the public domain. It is included within this chapter on that basis. It is attributed to JaGa.)

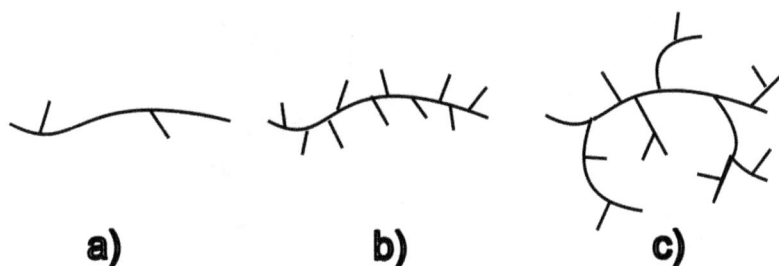

Figure 2.36. (a) High-density polyethylene with a low degree of short-chain branching, (b) linear-low-density polyethylene with a high degree of short-chain branching and (c) low-density polyethylene with a high degree of short and long-chain branching.

Table 2.7. Properties of polyethylene with different molecular structures, from [69].

Property/material	LDPE	LLDPE	HDPE
Density (g ml^{-1})	0.92	0.92–0.94	0.94–0.96
Melting temperature in °C	110	120–130	130–140
Tensile strength in MPa	24	37	43

figure 2.36(c)). The chain in between the two extremes, linear-low-density poly-ethylene (LLDPE), is the achieved by synthesizing copolymers with a small amount of α-olefins, for example α-butylene (figure 2.35).

The properties of the different kinds of PE are summarized in table 2.7.

The shared properties of all varieties of PE, such as low cost, lightweight nature, chemical resistance, and ease of processing, make them popular in a range of applications:

1. *Packaging materials and consumer products*: PE is used extensively in packaging materials and consumer products such as plastic bags, films, containers, bottles and kitchenware because of its durability, moisture and chemical resistance, cost-efficiency and ease of moulding (figure 2.37).
2. *Pipes and ducts*: Because of their excellent resistance to corrosion, chemicals, and weathering, HDPE and LLDPE are used for pipes and ducts.
3. *Electrical insulation for wires and cables*: All polyethylene is non-polar and therefore insulating. In particular HDPE is used as insulation for electrical wires and cables.
4. *Thermal and acoustic insulation*: As a foam, PE is used for thermal and acoustic insulation in the construction, packaging and automotive industry.
5. *Industrial applications*: PE is used as material for corrosion-resistant tanks, containers and trays, and in film form as liners or covers in agriculture, in all kinds of industry and food packaging.

Figure 2.37. Examples of PE bags with print.

2.3.1.13 Polypropylene (PP)

Polypropylene or less frequently polypropene is the polymer of propylene, first produced by G Natta in 1954 [70]. It comes in three configurations: isotatic, syndiotactic and atactic (figure 2.38).

Isotactic is the industrially relevant material. Industrial isotactiv PP has a melting point between 160 °C and 166 °C due to the presence of defects in the crystal structure. It is semi-rigid, tough, has good fatigue resistance, integral hinge property, good heat resistance, good chemical resistance and is translucent. It replaces PEs in the medium to high temperature range, for example warm water pipes or in medical/lab applications because it can withstand the temperatures of an autoclave [69]. It has a density of 0.905 g ml^{-1}, lower than all PEs, a melting point of 160 °C–166 °C, higher than all PEs, and a tensile strength of 33 MPa [71].

Its applications are [70]:

1. *Household goods*: Buckets, bowls, bottle crates, toys, bottle caps, bottles, kitchen appliances, hair dryers, irons, and luggage (figure 2.39).

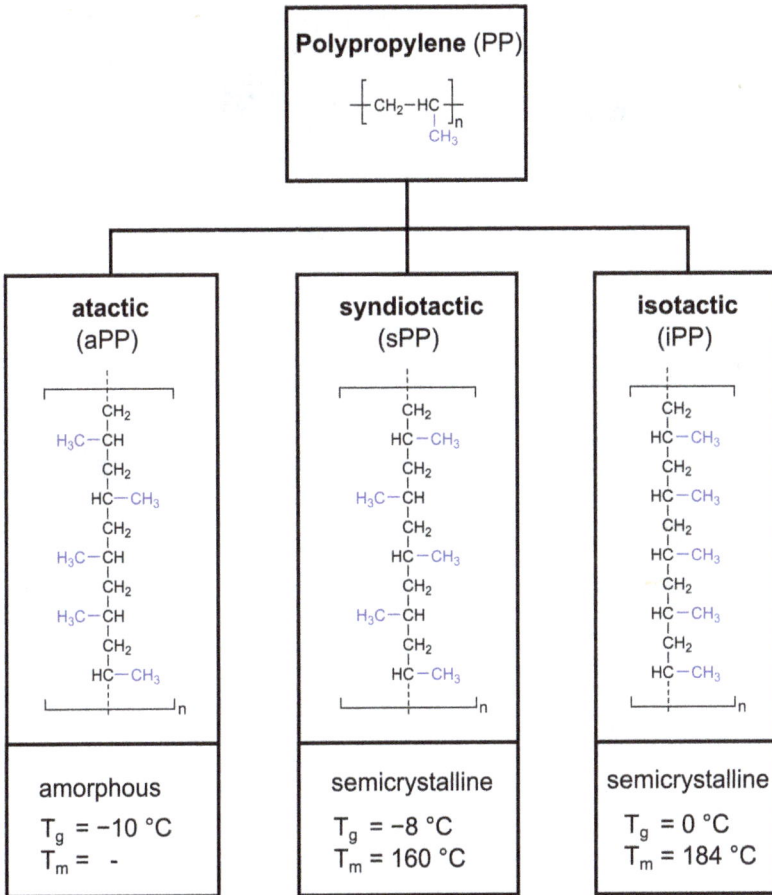

Figure 2.38. The three configurations of polypropylene. (This image has been obtained by the author from the Wikimedia website where it was made available by Minihaa (2008). It is stated to be in the public domain under a CC0.1 license. It is included within this chapter on that basis. It is attributed to Minihaa.)

2. *Domestic appliances*: Dishwasher parts, washing machine parts, refrigerator parts, and microwave parts.
3. *Automotive industry*: Radiator expansion tanks, brake fluid reservoir fittings, steering wheel covers, wheel arch liners, bumpers, side strips, and battery cases.
4. *Fibres*: Artificial sport surfaces, monofilaments for rope and cordage, stretched tapes, woven carpet backing, packaging sacks, and tarpaulins.
5. *Packaging*: Disposable food packaging, blister packaging, films, and strapping tape.
6. *Pipes and fittings*: Solid rods, punching plates, hot wire reservoirs, domestic wastewater pipes, pressure pipes, heat exchangers, corrugated pipes, and small diameter tubing.
7. *Furniture*: Stackable chairs.
8. *Medical and lab equipment*: Syringes, tubes, cartridges, and pipettes.

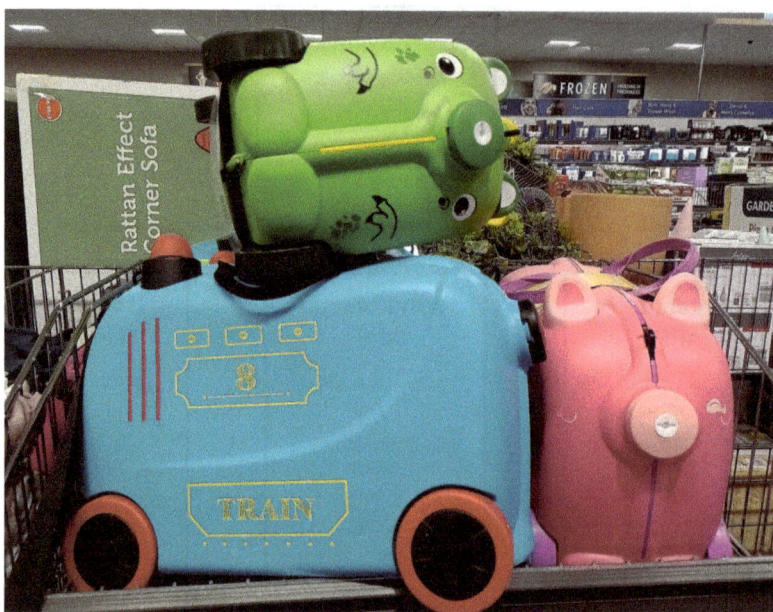

Figure 2.39. Propylene suitcases.

2.3.1.14 Polyvinyl chloride (PVC)

PVC is one of the most common plastics in the world. It accounts for 20% of all plastic production, second only to polyethylene [72]. In 1913 the first PVC patent was granted to the German chemist Heinrich August Klatte. In 1920 Waldon Semon, employee of the BF Goodrich company in Charlotte, NC, USA, added plasticizers that made PVC usable across many applications [73].

PVC is an amorphous polymer with a glass temperature of about 90 °C and is not a true plastic material [69]. It can be shaped when heated about 90 °C and cools down to a solid rigid material at room temperature. It has a density of 1.38 g ml^{-1} and a tensile strength of 2.6 MPa [72]. Because of its large chlorine content PVC is fire retardant. It is weatherproof and shows good chemical resistance.

PVC is manufactured from two starting materials, NaCl salt (57%) and 43% hydrocarbon feedstock, either crude oil or ethylene from sugar crops [72]. Polyvinyl chloride is polymerized by free radical polymerization by opening the double bond in the vinyl chloride monomer. The monomers arrange themselves into long, straight atactic chains with random positions of the chlorine atoms to minimize repulsion between them, see figure 2.40.

All commercial PVC has additives. Pure PVC is thermally unstable. Heat stabilizers prevent PVC from decomposing during moulding and also make it more resistant to daylight, weathering and heat aging. Added plasticizers makes PVC flexible, resilient and easier to handle. The most common plasticizers are low molecular weight phthalates (with restricted use in Europe) and high molecular weight phthalates (see figure 2.41).

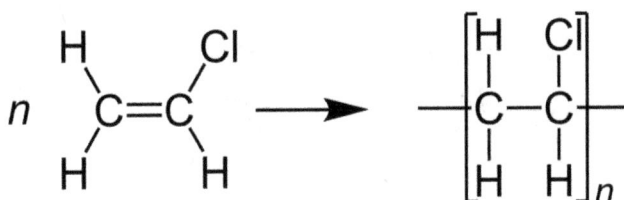

Figure 2.40. Polymerization mechanism of polyvinyl chloride. (This image has been obtained by the author from the Wikimedia website where it was made available by Jü (2020). It is stated to be in the public domain. It is included within this chapter on that basis. It is attributed to Jü.)

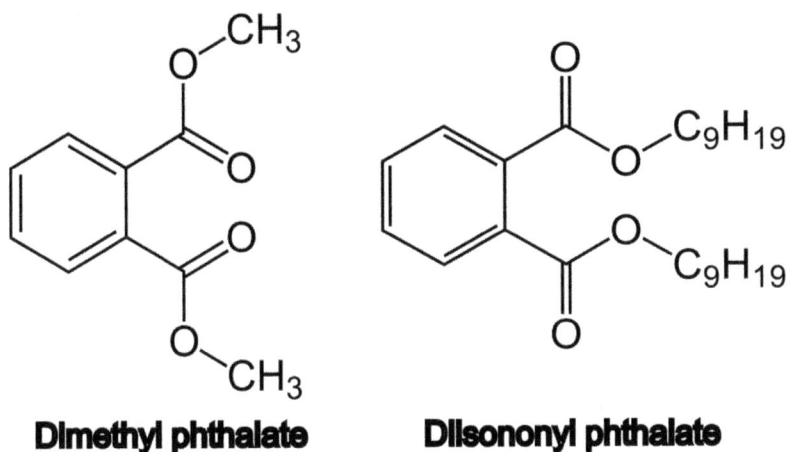

Dimethyl phthalate   **Diisononyl phthalate**

Figure 2.41. (a) Dimethyl phthalate, an example of a low molecular weight phthalate and (b) diisononyl phthalate, an example of a high molecular weight plasticiser. ((a) This image has been obtained by the author from the Wikimedia website where it was made available by Dschanz (2017). It is stated to be in the public domain. It is included within this chapter on that basis. It is attributed to Dschanz. (b) This image has been obtained by the author from the Wikimedia website where it was made available by Dschanz (2007). It is stated to be in the public domain. It is included within this chapter on that basis. It is attributed to Dschanz.)

Polyvinyl chloride is used in many different areas [72]:
1. *Construction*: Window and door profiles, conservatories, atria, pipes, fittings, cables, ducting, cladding, flooring, and wall covering.
2. *Healthcare*: 'Artificial skin' in emergency burns treatment, blood and plasma transfusion sets, blood vessels for artificial kidneys, catheters, blood bags, containers for intravenous solution administering sets, containers for urine continence and ostomy products, endotracheal tubing, feeding and pressure monitoring tubing, inhalation masks, surgical and examination gloves, shatter-proof bottles and jars, mattress and bedding covers, and blister and dosage packs for pharmaceuticals and medicines.
3. *Sports venues and sporting goods*: Performance sports surfaces, sports equipment, clothing, protective barriers, matting, wiring and piping infrastructure.

Figure 2.42. Example of a PVC banner.

4. *Coated fabric*: Shelters, tents, clothing, and bags.
5. *Electronics*: Cable insulation.
6. *Signage*: Signs, banners (figure 2.42).

2.3.1.15 Polystyrene (PS)

Polystyrene is the polymer of styrene, a synthetic monomer largely of fossil fuel origin. It was first described in 1839 by Eduard Simon who distilled styrene, he called styrol, from the oriental sweetgum tree *Liquidambar orientalis* [74]. It stayed a scientific curiosity until, in the 1930s, The Dow Chemical Company were the first to commercialize polystyrene successfully [75]. In 1948 they patented Styrofoam [76], the cellular foam well known from food and drink containers.

When styrene polymerizes the phenyl group can align in three configurations (see figure 2.43). Only atactic polystyrene is used commercially. Atactic polystyrene is amorphous, does not form a crystalline phase and is therefore not a true polymer. Below its glass transition temperature of 90 °C it behaves like a solid. At or above 90 °C, it becomes a viscous liquid and can be thermoformed [69]. As a high impact polystyrene, when it is hard and rigid, it has a density of 1.03 to 1.06 g ml^{-1} and a tensile strength of 2.2 to 2.7 MPa.

It is available as solid plastic (general purpose polystyrene (GPPS) and high impact polystyrene (HIPS)), rigid foam (expanded polystyrene (EPS) and extruded polystyrene (XPS)), and film (oriented polystyrene (OPS)).

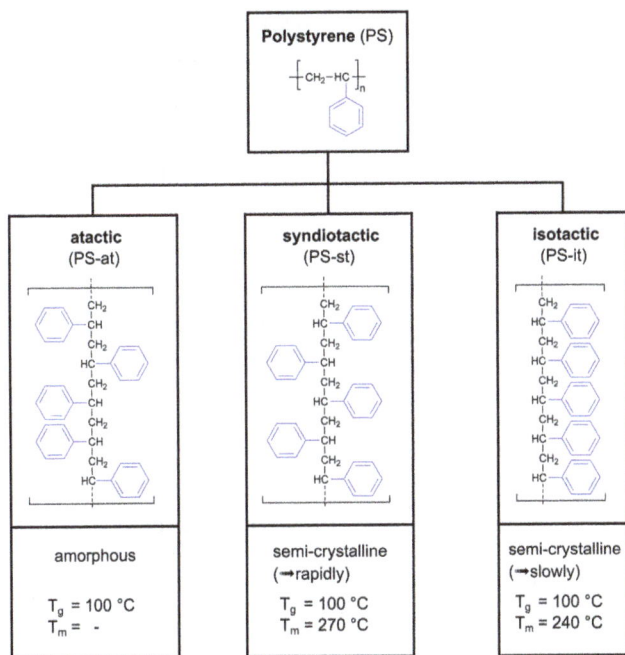

Figure 2.43. The three configurations of polystyrene. (This image has been obtained by the author from the Wikimedia website where it was made available by Minihaa (2019). It is stated to be in the public domain under a CC0 1.0 license. It is included within this chapter on that basis. It is attributed to Minihaa.)

Even though it is not recyclable, it finds many applications [77]:

1. *Packaging*: EPS (Styrofoam for example) is used to protect fragile items from impact during shipping. It is used in food packaging to keep food and drink warm, for example cups and take away containers. Solid HIPS is the material of vending cups, yogurt pots, meat trays, egg cartons and other single use food containers. It is used for packaging medical devices, pharmaceuticals, and other healthcare products.

2. *Construction*: EPS and XPS are used as thermal insulation in building walls, roofing and as decorative moulding.

3. *Electronics*: HIPS is used for the housing and other parts of IT equipment where form, function and aesthetics are combined, for example keyboards.

4. *White goods and kitchen appliances*: Solid polystyrene and polystyrene foam are used in refrigerators, microwave ovens, blenders, vacuum cleaners, air conditioning, etc.

5. *Disposable consumer goods*: Single use cutlery, plates and cups are used by fast food restaurants and for picnics and parties.

6. *Automotive*: Polystyrene is used in various automotive parts such as instrument panels, interior trim, and ventilation system components due to its lightweight nature and ease of moulding.

7. *Laboratory*: Polystyrene is used to make Petri dishes, test tubes, pipettes, and other lab equipment due to its transparency and chemical resistance.

2.3.1.16 Polyethylene terephthalate (PET)

According to Plastic Europe [78], 6.2% of the 320.7 Mt plastic produced in 2021 was PET. It was first patented by John Rex Whinfield and James Tennant Dickinson in 1941 in Great Britain, and then in 1945 in the USA [79]. PET is synthesized via polycondensation of terephthalic acid and ethylene glycol [80] (figure 2.44).

PET is most commonly used for the production of bottles for soft drinks. The PET bottle was patented by NC Wyeth and RN Roseveare in 1970 [81]. PET can exist in an amorphous form which is transparent or semi-crystalline. The semi-crystalline polymer is transparent when the crystallites are smaller than half the wavelength of light. It is white scattering when the crystallites are at least 500 nm or bigger.

PET is a versatile polymer with many applications because of its durability, clarity, strength and because it is easily recycled. The main applications are [82]:

1. *Packaging*: It is extensively used in the form of plastic bottles for carbonated water, juice and sauces (figure 2.45), in the form of containers and jars for food and cosmetics, and in the form of trays for vegetables and meat.
2. *Textiles*: As a fibre, then often just called polyester, it is used in clothing, upholstery and carpets, as filling for furniture and pillows. It can also be blended with other fibres to enhance moisture-wicking, stretchability, and wrinkle resistance.
3. *Film*: PET films (tradename Mylar) are used in food packaging, for electrical insulation, and medical equipment casing.
4. *Industrial applications*: Automobile tyre yarns, conveyor and drive belts, fire hoses, garden hoses, non-woven fabric for stabilizing drainage ditches, for nappies, disposable personal protective equipment (face masks, lab coats, etc).

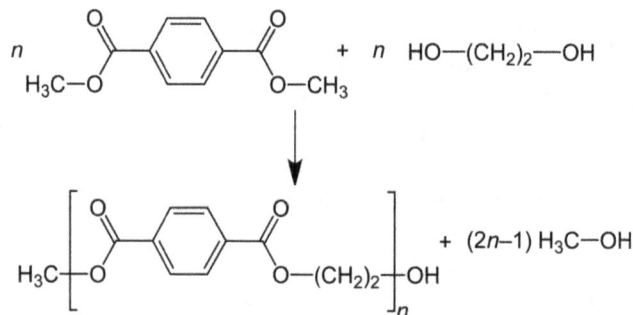

Figure 2.44. Polyesterification reaction in the synthesis of PET. (This PET image has been obtained by the author from the Wikimedia website where it was made available by Jü (2015), under a CC BY-SA 4.0 license. It is included within this chapter on that basis. It is attributed to Jü.)

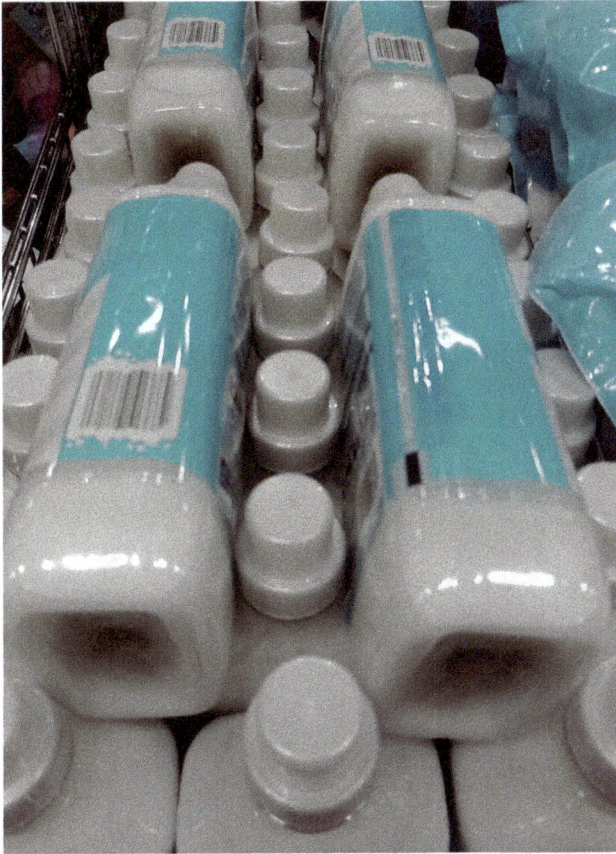

Figure 2.45. PET bottles.

2.3.1.17 Polyurethane (PU)

Polyurethane is a whole family of materials with different properties. It is formed through the reaction of isocyanates (compounds containing the functional group -N=C=) with compounds containing hydroxyl groups (-OH), such as polyols (figure 2.46).

The chemical reaction is called polyurethane synthesis and can result in a wide range of products with varying properties. Polyurethane was first synthesized by Otto Bayer and patented by IG Farben in 1937 [83, 84]. Since then its applications have expanded more and more [85]:

1. *Flexible foams*: They are used in upholstered furniture, mattresses, pillows, cushions and automotive seats. They provide comfort, support and shock absorption.
2. *Rigid foams*: Rigid PU foams are used in for insulating buildings, appliances, and refrigeration units due to their excellent thermal insulation properties. They are also used in packaging and buoyancy applications.
3. *Elastomers*: PU elastomers or PU rubber, are highly elastic and tough. They are used in various industries, including footwear, for example shoe soles

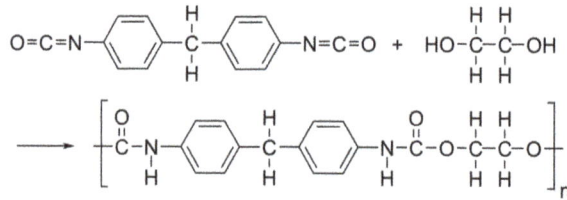

Figure 2.46. Polyurethane synthesis. (This image has been obtained by the author from the Wikimedia website where it was made available by Hbf878 (2017), under a CC BY-SA 3.0 license. It is included within this chapter on that basis. It is attributed to Hbf878.)

Figure 2.47. Polyurethane leather boots.

where they provide cushioning, support and resistance to wear and tear, all kinds of components for cars, for example bumpers, interior panels, gaskets, and seals, industrial belts, and medical devices, for example catheters, wound dressings, and prosthetic devices (figure 2.47).

4. *Coatings*: PU coatings provide protective and decorative finishes on surfaces such as wood, metal, and concrete. The offer resistance to chemicals, abrasion and weathering. They are applied to textiles to make them water resistance, breathable and durable.

5. *Adhesives and sealants*: Polyurethane-based adhesives and sealants are used in construction, automotive assembly, aerospace and other industries where strong bonding properties and flexibility is required.
6. *Sports equipment*: Either as foam or as elastomer, PU is used in the production of sports equipment, such as balls, padding, and protective gear due to its ability to absorb impact.

2.3.1.18 Nylon

Nylon was the outcome of industrial research conducted by DuPont between 1928 and 1934. In 1939 the first plant was opened in Seaford, DE, USA, producing nylon for stockings. A plaque unveiled in 1995 reads [86]:

At this site on December 15, 1939, DuPont began commercial production of nylon. Among the earliest successes of a fundamental research program novel in the American chemical industry, nylon was the first totally synthetic fiber to be fashioned into consumer products. Prepared wholly with materials readily derived from coal, air, and water, nylon has properties superior to its natural counterparts, such as silk. Nylon revolutionized the textile industry and led the way for a variety of synthetic materials that have had enormous social and economic impact on the fabric of everyday life worldwide.

Nylon is the trivial name for linear polyamides, either fully aliphatic or semi-aromatic. The preparation methods and the large number of starting monomers allows one to design macromolecules with task specific properties. Nylon polymers are named using numbers corresponding to the number of carbons between their acid and amine groups (figure 2.48), for example [69]:
- Homopolymers
  PA 6 [NH-(CH2)5-CO]n made from ε-caprolactam.
  PA 66 [NH–(CH2)6–NH–CO–(CH2)4–CO]n made from hexamethylene-diamine and adipic acid.
  PA 612 [NH–(CH2)6–NH–CO–(CH2)10–CO]n made from hexamethylenediamine and hexadecanedioic acid.
- Copolymers

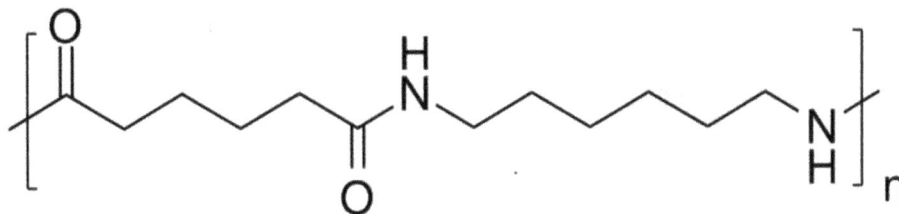

Figure 2.48. The chemical structure of nylon 46. (This image has been obtained by the author from the Wikimedia website where it was made available by Edgar181 (2018). It is stated to be in the public domain. It is included within this chapter on that basis. It is attributed to Edgar181.)

Table 2.8. Applications for modified nylons.

Modifier	Typical application
Unmodified	Cable ties
	Lighter bodies
Glass with or without mineral	Radiator end tanks
	Air intake manifolds
	Covers of various types
Mineral reinforced	Wheel covers
	Throttle bodies
Toughened	Clips
	Fasteners
	Ski bindings
Flame retarded	Swichgears
	Circuit breakers
	Other electrical/electronic components

PA 6/66 [NH–(CH2)6–NH–CO–(CH2)4–CO]n–[NH–(CH2)5–CO]m made from ε-caprolactam, hexamethylenediamine and adipic acid.

PA 66/610 [NH–(CH2)6–NH–CO–(CH2)4–CO]n–[NH–(CH2)6–NH–CO–(CH2)8–CO]m made from hexamethylenediamine, adipic acid and sebabic acid.

The repeat unit of PA, or nylon 66, was the first nylon ever and is still one of the most popular because of its combination of toughness, stiffness, high melting point and chemical resistance. It is mostly used for fibre production but modified for applications as listed in table 2.8 [139].

2.3.1.19 Acrylic (PMMA)
Acrylic covers a variety of polymeric materials obtained by polymerization of monomers with a vinyl groups to an acrylic monomer [69] (figure 2.49). PMMA or poly(methyl methacrylate) is the most common acrylic plastic, first commercialized in 1933 by Röhm and Haas AG in Germany as Plexiglas [87]. Because of its clarity and its UV fastness, it finds many applications as glass replacement:
1. *Architectural and construction*: Windows, doors, skylights, canopies, awnings, sound barriers at motorways or railways, and greenhouse panels.
2. *Automotive industry*: Head and taillights, and interior components such as dashboard panels.
3. *Signage and display*: Covers for illuminated signage, display cabinets in shops, shelving, and transparent signage.
4. *Furniture and home décor*: Tables, shelves, chairs, vases, decorative items, and sculptures.

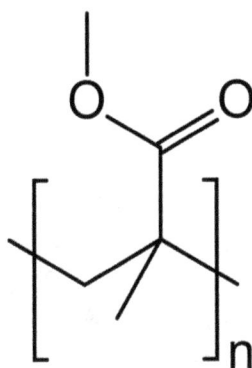

Figure 2.49. The skeletal formula of the polymethyl methacrylate repeating unit (Perspex, $(C_5O_2H_8)_n$). (This image has been obtained by the author from the Wikimedia website where it was made available by Dr Torsten Henning (2008). It is stated to be in the public domain. It is included within this chapter on that basis. It is attributed to Dr Torsten Henning.)

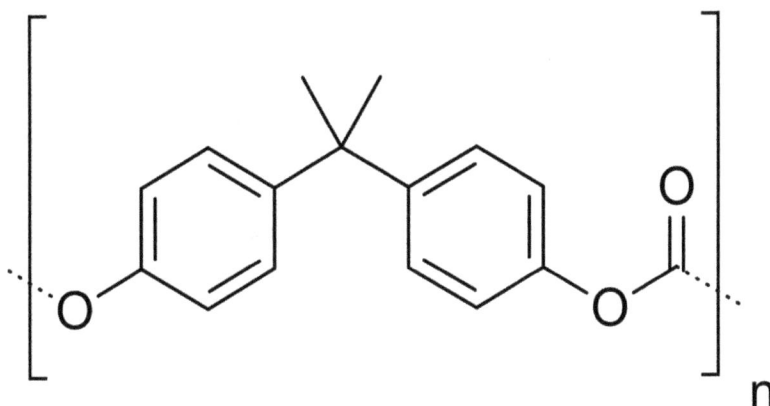

Figure 2.50. Repeating chemical structure unit of polycarbonate made from bisphenol A. (This image has been obtained by the author from the Wikimedia website where it was made available by Edgar181 (2016). It is stated to be in the public domain. It is included within this chapter on that basis. It is attributed to Edgar181.)

5. *Cameras and electronics*: Optical lenses for phones, cameras, projectors, and protective screens for smartphones and tablets.
6. *Aviation*: Aircraft windows and canopies.
7. *Medical and laboratory equipment*: Intraocular prosthesis, syringes, luer tapers, test kits, blood filters, flowmeters, incubators, surgical trays, etc.

2.3.1.20 Polycarbonate (PC)

Polycarbonates are formed by carbonate links connecting hydrocarbon groups (figure 2.50). If the hydrocarbon group is an aromatic group, the polymer cannot crystalize, has a glass transition temperature above 100 °C, is transparent, and is

Figure 2.51. Acrylic frame with polycarbonate lenses.

strong and tough. In 1953 Hermann Schnell, employee of Bayer, patented the first linear polycarbonate [88]. Many patents have followed.

Polycarbonate has similar features as poly(methyl methacrylate) and similar applications, but is stronger, more scratch resistant and can withstand higher temperatures. It is used for:

1. *Eyewear*: Sunglasses, prescription lenses, safety goggles, and face shields (figure 2.51).
2. *Packaging*: Bottles and containers.
3. *Electronics*: Casings for laptops and mobile phones.
4. *Automotive*: Head and taillights, panoramic roofs.
5. *Construction*: Windows, skylights, canopies, and greenhouses.
6. *Aviation*: Windows and canopies.
7. *Medical and laboratory*: Carboys, bottles, flasks, culture vessels, filtration apparatus, jars, dessicators, cryogenic storage, centrifugeware, and safety shields.

2.3.1.21 *Printing on plastic*

As a substrate, plastic is as versatile as paper, but intaglio printing methods are usually not used on plastic. Here we will not discuss printing on plastic fibres, woven or non-woven. We will discuss printing on non-porous plastic. It poses certain difficulties:

1. *Adhesion*: Plastic surfaces are often non-porous and have low surface energy, making it difficult for inks to adhere which can result in poor print quality, smudging, or the ink can be easily scratched off after drying. In figure 2.52 average values for different surface energies are displayed. Paper is not included because its surface energy depends very much on the sizing and surface treatment. The higher the surface energy, the better the ink spreads

surface energy in mN/m vs Material

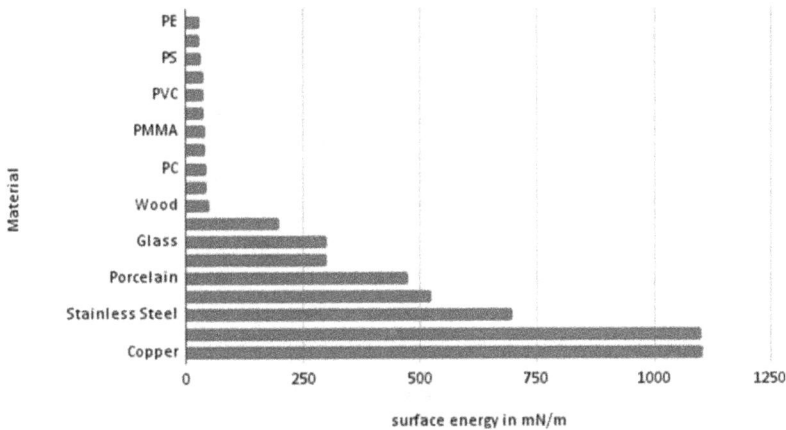

Figure 2.52. Surface energy for different materials, collected from different sources on the Internet [89, 90].

and the easier it bonds to the surface. The lower the surface energy, the less wetting occurs. The surface becomes non-stick. All the hydrocarbon polymers discussed in this chapter are in the low surface energy range. In general inks suitable for printing on paper are not suitable for printing on plastic. Since the plastic is non-porous the ink will not penetrate the surface and inks which shrink during drying substantially, i.e. inks with a high solvent content, will fall off after drying. Inks for printing on plastic must be especially formulated and surfaces must be treated before any printing commences. The best option is often UV curable ink.

2. *Drying and curing*: We touched already on a problem with drying: shrinkage. Another one is deformation of the substrate when it is exposed to heat. Depending on the glass transition temperature and the heat required to cure ink, the ink may not be compatible with the plastic type because it will become soft and deform. Unlike with paper, it is not easily remedied.

3. *Static electricity*: Unlike any other substrates discussed in this book, except synthetic textiles, plastic will quickly acquire charge via friction. Static electricity must be tightly controlled because the handling of the substrate becomes difficult. When charged it attracts dust from the air which will deteriorate the print quality. It can even deflect droplets when inkjet printing is used. Charge changes the surface energy and ink deposition becomes unpredictable. Static electricity can make the plastic stick to machinery or itself and can cause an electric shock when touched. Sparks from charged plastic sheets are a fire hazard when solvent based inks are used. It poses a health and safety hazard.

4. *Deinking*: Plastic packaging is heavily printed. Contamination by printing inks is one of the main reasons why flexible packing is difficult to recycle. Residual ink changes the colour of the polymer, makes films less stiff, weaker and denser

than the original material, and its decomposition can produce gas generating a rancid odour [91]. In general, plastic with print can only be downcycled.

Otherwise printing on plastic is similar to printing on paper. Plastic as paper comes in many forms, from soft to rigid, from porous to sealed. The application dictates the printing process and the inks which can be used. Today, in 2023, plastic has lost its innovative reputation. It is seen as one of the main culprits when it comes to environmental pollution. There is a big push to replace it with other, renewable, materials, such as paper, but it production and its functionality make it cheap and fit for many tasks. It will not disappear so quickly from our lives.

2.4 Printing on textiles

Paper and textiles can be made out of the same materials. '[Paper] is a sheet material made up of a network of natural cellulosic fibres which have been deposited from an aqueous suspension' [10]. Textiles can be constructed in different ways and from many different materials, cellulose, such as cotton, protein, such as wool and silk, and man-made (see table 2.13), such as polyester, for example. The construction of fabric by weaving, to mention only one possibility, generates a highly structured surface which is taken to an extreme in terrycloth, a fabric for bath towels woven with a loop or pile on special looms called dobby looms. The bath towel is a good example for the handling that printed fabric is exposed to when it is in the hands of the end user. A towel is made to absorb water from a surface. The print should not dissolve in water or rub off on the surface it comes in contact with. Towels are washed in warm or hot water. The print has to be resistant to chemical agents such as surfactants, an ingredient in washing powders. When the towel is dried on a washing line or used on the beach, it is exposed to UV radiation which can destroy colourants. The print has to be light fast. A print on textile has to survive a lot of interactions with its environment for long periods of time.

2.4.1 The history of printing on textiles

Even though 'print' today is mostly associated with print on paper or cardboard, print on textiles predates print on paper. India might be the origin of all textile printing. Even though no printing blocks or textiles have survived, it is believed that textiles were being printed by the Indus Valley Civilization as early as 3000 BC [92]. Finds at the site of Mohenjodaro in present day Pakistan are dated between 2600 and 1300 BCE and provide evidence of cotton processing and dying, but no clear evidence of printing could be established [93]. Textiles are biodegradable and survive only under extraordinary circumstances, as in the burial mounds and peat bogs of Denmark.

Complete costume sets have been preserved in the Bronze Age graves at Borum Eshøj, Egtved, Skyrdstrup, Trindhøj, Guldhøj and Muldbjerg dated to the fourteenth to twelfth century BC [94]. Due to the tannin-rich and wet environment in peat bogs, all textile finds have lost their original coloration and now present different shades of brown. For a while, it was assumed that only the natural shades of wool were used to generated patterns, but spectroscopic and chemical analysis confirmed that the garments were actually very colourful and could have been elaborately decorated [95–97]. The traces of dyes found belong to the three main groups of natural colourants: anthraquinone, flavonoid and indigoid dyes [96]. The most important natural anthraquinone dye is alizarin which can be extracted from dyer's madder (*Rubia tinctorum*).

Other sources for dyestuff of the anthraquinone group, such as carminic acid, kermesic acid and laccainic acid, are scale insects, such as cochineals and kermes and the secretion of lac insects. Flavonoid dyes are yellow. Sources are weld (*Reseda luteola* L.), sawwort (*Serratula tinctoria* L.), dyer's broom (*Genista tinctoria* L.) or chamomile types (*Anthemis* species). Indigoid dyes are the most important blue dyes. In Europe they were extracted from woad (*Isatis tinctoria* L.) and in Asia, South America and Africa from indigo (*Indigofera* or *Polygonum* species). It is almost unthinkable that simple forms of stencil printing would not been employed much earlier, but the first solid evidence of printed textiles is from a textile recovered in a Greco-Scythian burial in Crimea, which dates to the fifth–fourth century BCE [98] and a tunic excavated by Robert Forrer in Akhmin, Egypt, dated to the sixth–seventh century (see figure 2.53). The burial fields near Akhmin in upper Egypt were first excavated in 1885/1886 [99] and yielded a treasure trove of textiles. The burial ground was used over several centuries, from the Greco-Roman era through Byzantian times to the Arab Conquest [100]. No archaeological documentation of the site exists, but the description in [101] by Robert Forrer allows an approximate location of the burial fields and the type of burials found there. When Forrer arrived in 1894 the site had been known by archaeologists and grave robbers for ten years and had been exploited for scholarly and financial reasons. The excavations had left a field of devastation: 'Wahrscheinlich kein Anblick fuer zartnervige Leute, ein

Figure 2.53. Dye resist printing on a child's tunic, cotton or linen. Excavated by R Forrer at Akhmim in 1894; in the collection of the Victoria and Albert Museum. Not on display. (Copyright Victoria and Albert Museum, London.)

Schlachtfeldbild ergreifendster Art' ('Probably not a sight for delicate people, a poignant view of a battle field') [101]. When the site was first excavated, information of where and how was not recorded, the ornamental parts were cut of the complete textile and neither the original garment nor a record of it was kept. Even though textiles from Akhmim can be found in almost all archaeological collections, their dating and interpretation is problematic.

The oldest extant Indian block prints come again from Egypt, fragments dating from the ninth to the seventh centuries, found at various sites including al-Fustat near Cairo, Quseir-al-Qadim, Berenike at the Red Sea, and Qasr Ibrim and Gebel Adda, both in Nubia, southern Egypt, demonstrating the trade links between India and the Mediterranean [93]. Many of the fragments combine block printing with hand painting. Towards the East, Indian textiles were the currency of the spice trade. Arabic, Chinese and European sources all speak of the dominance of India as source of splendidly coloured silks and cottons, light and wash-fast [102–104]. India dominated the Indian Ocean textile trading network, spanning from Cairo to Tokyo until the arrival of the first western venturers in search of direct access to the spice trade in the fifteenth century [93]. In 1453 the Ottomans captured Constantinople and disrupted the Middle Eastern trading routes. The Portuguese were the first to establish themselves in South East Asia. Vasco da Gama reached Calicut, modern Kozhikode, Kerala on the Malabar Coast in 1498. In 1510 Alfonso de Albuquerque invaded Goa and established the Estado da India. In 1521 Ferdinand Magellan's expedition funded by Spain reached the Moluccas [93]. The English, French, and Dutch established their trading companies in the seventh century: The English East India Company was founded in 1600, the Dutch Vereenigde Oost—Indische Compagnie in 1602 and the French Compagnie des Indes Orientales in 1664. The Portuguese were the first to disrupt the Asian/Arab trade structure, but by the seventeenth century they had to yield to the Dutch who became dominant in the Indonesian archipelago. The English concentrated on mainland India after they were defeated by the Dutch in 1618 in the Gulf of Siam. Both companies established headquarters and manufacturing sites in India and South Asia, producing products not only for the spice trade, but also for the cloth European market. The uptake in Europe was slow at first, but with the advent of a hybrid style of chintz comprising Chinese, European and Indian influences and with an aggressive marketing strategy and 'brand ambassadors', the English East India company started a boom in sales to fashionable English society.

By 1664 textiles accounted for about 75% of all exported goods from India [93]. The import of Indian cotton into Europe caused alarm. The European manufacturers of woollen, linen and silk fabric felt under threat. The governments intervened. In 1686 France banned the import of chintz to protect its silk industry. In 1697 wool and silk weavers stormed the headquarters of the English East India Company in London and in 1719 weavers doused women wearing calico in nitric acid (acid attacks have a long tradition), tore the clothing off their bodies and marched on the calico-printing workshops in Lewisham [105]. The 1701 Calico Act ('An Act for the more effectual employing the Poor, by encouraging the Manufactures of this Kingdom'), reinforced in 1720, prohibited the retail and

Figure 2.54. Toile de Jouy, depicting the handling of the printed fabric. (Photograph taken in the Musee d'art et d'histoire in Orange, France.)

consumption of Indian textiles in the United Kingdom, but not in the American colonies. The textiles could be imported and then exported to North America which influences fashion and interior design in the colonies. In the second half of the eighteenth century the ban was relaxed all over Europe, but the tide had turned. Technical innovations in Europe brought an end to the world dominance of Indian block print. The production of Toile de Jouy is an example of the industrial power developing in Europe from 1750 onwards (figure 2.54).

In 1756 Francis Nixon from Drumcondra, near Dublin, Ireland brought copper plate printing to Merton in Surrey, England, where he set up works for textile printing. In 1760 Christophe–Philippe Oberkampf founded his printing workshop in Jouy, France, which would give the fabric its name. In 1782, he developed a method to make the print wash-fast. The colours were derived from madder for black, red enhanced by Brazilwood, purple and brown, indigo for blue, weld for yellow, orchil, a lichen from the Canary Islands, for violet and scarlet from cochineal [92]. The printers used these dyes in a trial-and-error method which led, in the case of Toile de Jouy, to exquisite results. The first step of the printing process was to bleach the fabric in open air to give a white background (in Germany it is said that moonlight gives the whitest white). Several printing processes were used, either on their own or in combination to produce the finished cloth: (i) *direct print*; (ii) *discharge print*: the fabric was dyed and the pattern was bleached out of the coloured background; (iii) a *mordant* was printed on the fabric and the cloth was then dipped into a dye bath, and the dye bath dyed only the mordant; and (iv) *resist printing*: wax or a starch paste was printed upon the cloth before it was immersed into the dye bath. The mordant and resist method came from India and were the most important textile printing methods until the advent of synthetic dyes.

In 1783 Thomas Bell patented the first successful rotary printing machine. Since the beginning of the nineteenth century, Oberkampf had rotary printing machines in his workshop. At the beginning the design was based on Indian samples, but soon it developed into a new direction [106]. The copper plate technique and mechanical roller printing allowed more complicated designs with a much higher production output than could be achieved by block printing.

From the beginning of the nineteenth century European chemists started adding new dyestuff to the colour palette. In 1834 Friedlieb Ferdinand Runge made 'kyanol' from coal tar. With the accidental discovery of mauvine, an aniline dye, by William Henry Perkin in 1856, and the rapid development of further synthetic dyes, Europe had found its own source of cheap dyes for dying and printing. In parallel with the rise of production power in Europe, the political landscape on the Indian subcontinent shifted. By two of Acts of Parliament (1173 and 1784) the character of the East India Company was change from a trading company to a representative of the English crown on the Indian subcontinent [107]. In 1858 the British Crown assumed direct control and in 1877 Queen Victoria was proclaimed Empress of India. The Indian textile market became an expansion ground for British textile printing. In the *Colonial and Indian Exhibition, 1886. Empire of India* descriptive catalogue by Thomas Wardle [108], himself a prominent British silk dyer and printer, a comment on page 148 shows the commercial pressure textile printing in India was exposed to:

> Every village has its dyer if not its calico-printer, and the people are fond of coloured garments. But aniline colours are rapidly taking the place of the more expensive but permanent vegetable dyes, just as European steam calico-printing is steadily carrying ruin to the homes of the Indian calico-printer.

The industrial revolution enabled manufactures of printed textiles to turn out many thousands of metres of cheap and colourful designs in the same time and at less cost than one length of block printed cloth produced by hand. The industrial mass-produced ware did not only provoked regret, but also objections. In England John Ruskin objected to the working conditions 'Alas! if read rightly, these perfectnesses are signs of a slavery in our England a thousand times more bitter and more degrading than that of the scourged African, or helot Greek' [109] and William Morris was appalled by the crude design, so much that he demonstrated his dissatisfaction with a 'sit-in' at the Great Exhibition organized by the Royal Society of Arts in 1851. His and his company motto became: 'Art made by the people, and for the people, as a happiness to the maker and user.' [110]. He returned to block printing and rejected the use of synthetic dyes, he used the chintz-palette of the early Indian cotton printers: the blue of indigo or woad, the red of madder, the yellow of weld or Persian berry, the brown of walnut juice (getting his green and purple and black by combinations of these) embracing the transient nature of the dyestuff, 'They are not eternal; the Sun, in lighting them and beautifying them, consumes them, yet gradually and for the most part kindly' [110]. In working conditions and design, he defined the 'Art and Craft movement'. His designs are still in circulation, see figure 2.55.

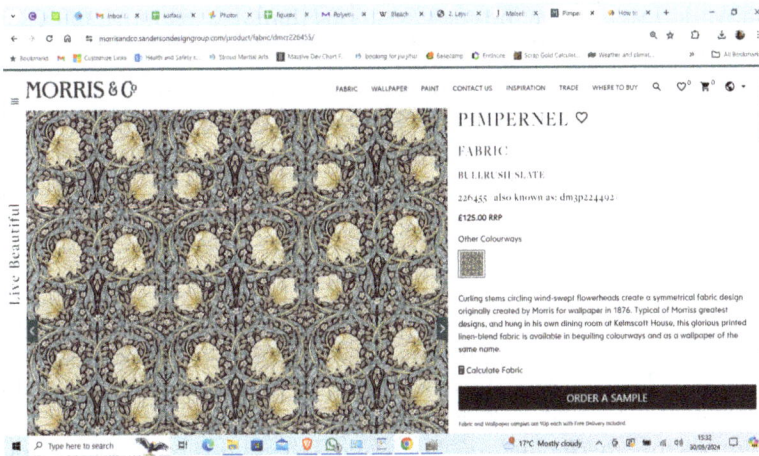

Figure 2.55. Screenshot of a webpage of one of the many companies still selling William Morris designs. (Screenshot reproduced with permission from Morris & Co. Copyright Morris & Co.)

Art Nouveau (or Modern Style in Britain) is the successor to the Art and Crafts movement and had the same objective—to break down the distinction between fine art and applied art—but embraced industrial production methods and was forward-looking instead of romanticizing the past. The Exposition Universelle in Paris in 1900 presented the work of designers whose names are still known today: Lalique, Daum, Gallé, Bugatti, and Tiffany. Liberty & Co became known for its Art Nouveau textiles in Europe and especially in France. Their style was so recognizable that it was called *le style Liberty* in France or *stile Liberty* in Italy. Originally the print works were at Merton Abbey, using block print for their avant-garde textile design, today the print works are operated near Lake Como in Italy using screen print for their huge library of 45 000 designs dating from the 1800s to the present day [111]. Even though the Bauhaus, the German offshoot of the Arts and Crafts movement revolutionizing design between the wars, did not teach or produce printed textiles, its teachings affected all crafts in Europe during the Art Deco period of the twenties and thirties. The influence of Walter Gropius and Johannes Itten on printed textiles manifested itself in patterns of simple geometric forms—cubes, spheres, and triangles [92].

In the 1930s high throughput textile printing done with copper roller print machines was gradually replaced by screen printing, first in the middle and upper price ranges, and 20 years later in all price ranges. Screen printing saved the market for printed textiles from complete collapse. The Great Depression (1929–39) caused a shrinkage of all markets. At the same time textile designers lost touch with the wider consumer base and the demand for new fabrics declined, less fabric was sold and because of the initial cost for every new design the price per metre of fabric went up and up, becoming more and more unaffordable. Bankruptcy as a consequence of shrinking markets and excessive costs was prevented by the advent of screen printing. The Second World War and its immediate aftermath were difficult for all types of textile printing. Textile printing was not important for the war effort and design innovation went more into new weapons than new patterns. Only after 1947 did the industry catch up with new machinery and started again with commissioning designs from leading artists [92]. Synthetic fibres, a new generation of synthetic dyes, the increased production speed of screen printing and its capability to carry out much bigger designs than roller printing, led to fast development and an almost unlimited number of styles. In the fast moving world of fashion, four distinct classes of textiles developed [92]:

1. *High-end*, catering to the avant-garde customer with money, bespoke interior soft furnishings, and haute couture.
2. *Boutique market*, catering for the individualistic customer who searches for non-mainstream design, but does not have same the budget as the high-end customer. The market is the entrance level for young designers and includes the print-on-demand market and the upcycling market.
3. *Conservative interior and soft furnishing*, represented by Sanderson or Liberty fabrics.
4. *Bread and butter fabrics*, today represented for example by Ikea for soft furnishing and interiors.

The ever-increasing and instant availability of screen printed fabric is allowing rapid changes of style and colour. Already in 1969 it was stated that the times of a few fashion shows setting the trend for the season are over [92]. Today, print-on-demand, via sublimation, digital or screen print, has made customization possible to an extent that printed fabrics can be one-offs and trends can change weekly.

The increased demand for unique clothing and easy access via online shops is expected to have a positive impact on market growth. In 2018 the printed textile market was estimated at $146.5 billion and the compound annual growth rate (CAGR) is predicted as 8.9% until the year 2025 [112]. Digital textile printing is growing rapidly but still only has a small market share. In 2019 the global digital printing market was valued at $2.2 billion, just 1.4% of the global printed textile market, but its CAGR is predicted at 19.1%, double the CAGR of the market in general. In 2027 it will have doubled its global market share, but traditional printing will still account for 97% of the commercial market (figure 2.56).

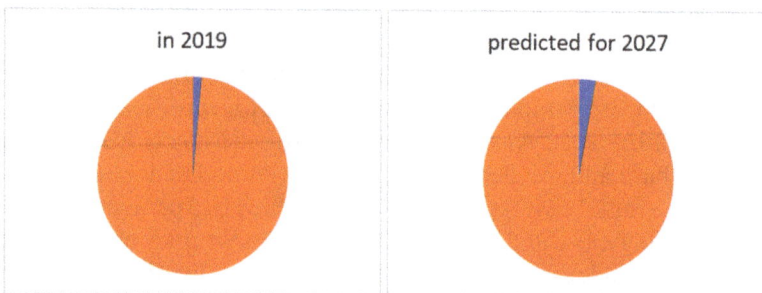

Figure 2.56. Percentage of digital textile printing (blue) of the total global printed textile market (orange). On the left side is the value from 2019 and on the right the predicted values for 2027.

Natural fibre in 1000 metric tons and Synthetic fibre in 1000 metric tons

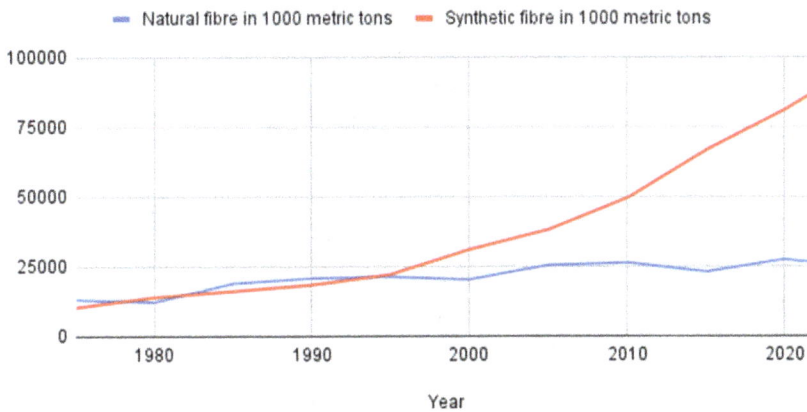

Figure 2.57. Comparison of the production of natural fibres and synthetic fibres from 1975 to 2022. (Based on data published by www.statista.com.)

Printing techniques, such as rotary screen printing, automated flat screen printing, hand screen printing and dye sublimation transfer printing dominate the mass production of printed textiles. On the one hand, in screen printing any ink can be used independent of the substrate, on the other hand, this process is coming under pressure because of its low sustainability. Digital technology belongs to a more sustainable, although still niche, market which includes block, roller, and batik printing. Textiles for printing have undergone a substantial change in the last 30 years. Until 2000 fibres were mostly natural, after 2000 most fibres have been man-made, see figure 2.57.

Cotton is still the most popular substrate for textile printing. It is durable, biodegradable, absorbing and insulating, ideal for apparel and household textiles, but its production is water hungry and unsustainable. Polyester has not yet obtained a good reputation but is much more environmentally friendly in some respects, although not in others. Its low absorbance causes problems for traditional printing except for dye sublimation, therefore hybrid or blend fabrics are the better choice for print. An increasing demand for printed silk in the Gulf region and Japan will lead to a growth in these markets [112].

The construction of fabric

Independent of the material of the fibre, textiles are made using the following methods:

- *Weaving*: An array of threads (wrap) is put under tension on a frame, the loom. A second thread, weft, is then inserted into the first array, passing alternately over and under the first set of threads, see figure 2.58. Depending on the stiffness, tension and spacing of the wrap and weft, the fabric can be stiff or soft and flowing. The elasticity of the fabric depends on the elasticity of the yarn and is not a result of the weaving process.

Figure 2.58. Example of plain weave. The red thread is the warp and the blue is the weft. (This image has been obtained by the author from the Wikimedia website where it was made available by Alfred Barlow, Ryj, PKM (2012), under a CC BY-SA 3.0 license. It is included within this chapter on that basis. It is attributed to Alfred Barlow, Ryj, PKM.)

- *Knitting*: A continuous thread is manipulated by two needles creating connected loops. The number of loops is determined by the number of active loops which are transferred from one needle to the other during the process of knitting, see figure 2.59. The process of knitting creates an elastic fabric with a stretch up to five times its original dimensions and is often used for shape hugging items.
- *Crochet*: Crochet is another technique based on loops. Only one needle is used and only one stitch is active, see figure 2.60 The fabric created by crochet is stiffer compared to that created by knitting and the surface is much more structured than that created by knitting. Crochet fabric is not used as a substrate for printing.

Figure 2.59. An example of knitting in action. The stitches on the needles are active.

Figure 2.60. Example of crochet. Only one stitch is active.

- *Lace making*: Lace is an open, web-like fabric. It can be made by knitting, crocheting, needle work or bobbins on a cushion. Lace is not suitable for printing since the fabric does not form a continuous substrate (figure 2.61).
- *Macramé*: The fabric is generated by knotting and a prominent surface structure is generated that is unsuitable for print (figure 2.62).

Figure 2.61. Example of lace made on a cushion with bobbins. Image by Susanne Klein.

- *Felting*: Felt is very similar to paper. It is a non-woven fabric, see figure 2.63. Hot water is applied to animal hair (see table 2.11) and the wet animal hair is agitated which makes the hair entangle and form a dense fabric (an undesirable example is the shrinking of a woollen jumper when washed in washing machine at too high a temperature and with too much agitation). Felt can be made from animal hair, acrylic, acrylonitrile or rayon.

All begins with a fibre. A fibre is a long, thin and flexible structure which can be of animal, plant, mineral or hydrocarbon origin.

In table 2.9 all the plant fibres which are used for fabric production are listed.

2.4.1.1 Preparation of fabrics for printing

An important difference between printing on paper and printing on fabric is that paper arrives at the printers in a ready-to-use state, whereas fabric needs to be prepared for printing. A fabric is said to be in the grey or loom state (for woven fabrics) when it has not been processed further after it has left the knitting machine or the loom [127]. Depending on the fibre, impurities have to be removed (scouring), the surface has to be smoothed, the colour of the substrate has to be adjusted by bleaching, and pre-treatments have to be applied.

For natural and regenerated cellulosic fibres (see table 2.9 and the first four entries in table 2.10), unless the fibre in the fabric is uniform in whiteness, absorbency and chemical composition, and has low levels of contaminants (wax, lignin, electrolytes, vegetable debris, etc), the result of any printing process will be unsatisfactory (table 2.12). A successful pre-treatment, i.e. an even print, depends on several factors [133]:

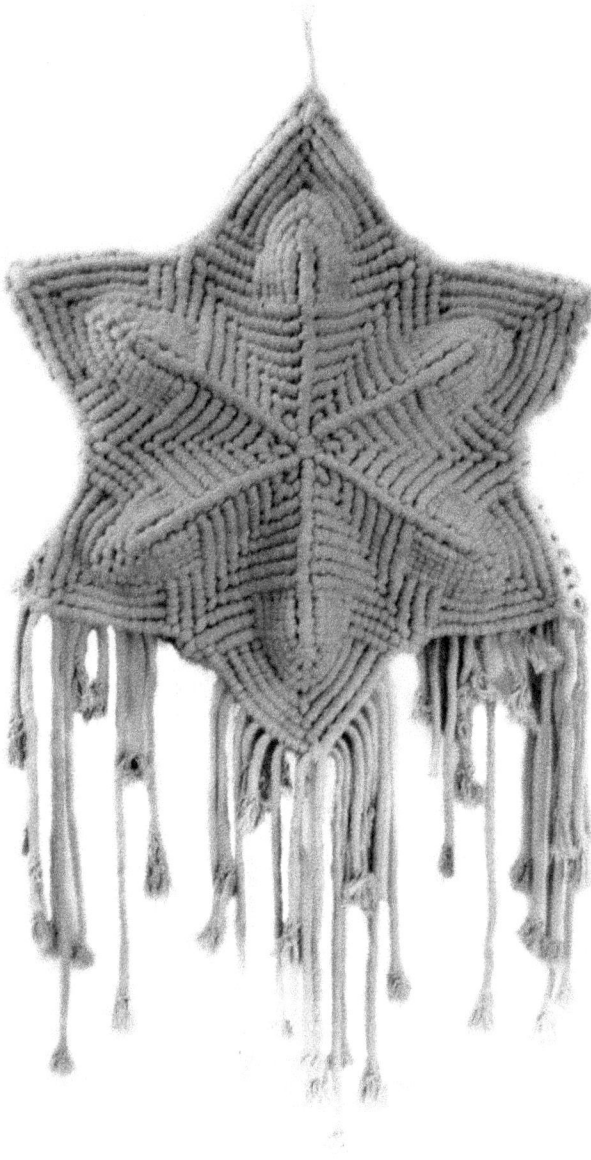

Figure 2.62. Example of macramé fabric. (This image has been obtained by the author from the Wikimedia website where it was made available by Wupo (2016) under a CC BY SA 4.0 licence. It is included within this chapter on that basis. It is attributed to Wupo.)

- The level and type of impurities.
- The chemicals used in the process.
- The water supply.
- The machinery used in the process.

Figure 2.63. Example of industrial felt.

Most natural cellulosic fibres contain vegetable debris produced by mechanical harvesting, hemicellulose, pectins, lignin, proteins, wax and ash. The impurities in regenerated cellulosic fibres generally come from the solvents used to dissolve the cellulose. Further impurities are added during the processing of the fibres into yarns and then the final fabric, such as size, spin finish and knitting lubricant [133].

With a 25% market share in global fibre production in 2019, cotton is the second most important fibre after polyester (58%) [134]. The preparation of woven cotton fabrics for printing consists of three stages—desizing, scouring and bleaching—, and sometimes singeing. For knitted fabric desizing is often not necessary [133].

  1. *Singeing*

     Many woven cotton fabrics are singed first to remove excess surface fibre. The fabric passes through a brushing unit which lifts loose surface fibre and is then passes at high speed through gas fired burners where both sides of the fabric are singed. To quench and sparks, the singed fabric passes through steam or liquor.

  2. *Desizing*

     To smoothen the warp yarn and increase the speed the weft can be inserted with, yarns are sized with the following materials:
     - Starch from potatoes, maize, rice or tapioca.
     - Starch ethers or esters.
     - Organic polymers (polyacrylates, carboxymethylcellulose, methyl-cellulose, polyester, poly(vinyl alcohol).

Table 2.9. A list of plant fibres used for fabric production. (The cellulose image has been obtained by the author from the Wikimedia website where it was made available by CeresVesta (2009). It is stated to be in the public domain. It is included within this chapter on that basis. It is attributed to CeresVesta.)

Name	Origin of fibre	Hard fibre	Soft or bast fibre	Seed fibre	Length of fibre	Diameter of fibre
Plant fibres (cellulose based)	Structural formula of cellulose					
Flax [113, 114]	From the single stem of *Linum usitatissimum*		x		20–130 cm	40–620 μm
Hemp [114, 115]	From the stem of *Cannabis sativa* L.		x		1–5 m	16–50 μm
Jute or Hessian [114, 116]	From the inner bark of *Corchorus capsularis* or *Corchorus olitorius*		x		2.4–4 m	30–140 μm
Nettle [117]	From the stem of *Urtica dioica* L.		x		27–73 mm	10–63 μm
Ramie [114, 118, 119]	From the bark and the stem of *Boehmeria nivea*		x		Up to 1.9 m	40–60 μm
Kenaf [114]	From the bark and the stem of *Hibiscus cannabinus* L.		x		Up to 3 m	40–90 μm
Banana [114, 120]	From the stem of banana or plantain plants		x		2.5–13 mm	50–280 μm
Sisal [114, 119]	From the leave of *Agave sisalana*	x			Up to 1 m	100–400 μm
Abaca [114, 121]	From the leaf sheath of the *Musa textilis*	x			1.8–3.7 m	17–21 μm
Piña [114, 122]	From the leaf of pineapple	x			0.5–1 m	200–8800 μm
Raffia [123, 124]	From the leaf of the *Raphia farinifera*	x			Up to 1.5 m	1.53 ± 0.29 mm
Cotton [114, 124]	From the exterior of the seeds of *Gossypium*			x	1.6–6 cm	16–21 μm
Coir [125]	From the husk of *Cocos nucifera*	x			44–305 mm	100–795 μm
Kapok [124]	From the seeds of *Ceiba pentandra*			x	20–30mm	20 μm
Bamboo [126]	From the whole stem of *Phyllostachys* or *Ochlandra travancorica*		x		1–4 mm	10–30 μm

Table 2.10. Table of natural polymers. (Cellulose acetate: this image has been obtained by the author from the Wikimedia website where it was made available by Jü (2013). It is stated to be in the public domain. It is included within this chapter on that basis It is attributed to Jü. Cellulose triacetate: this image has been obtained by the author from the Wikimedia website where it was made available by Edgar181 (2008). It is included within this chapter on that basis. It is attributed to Edgar181. Corn fibre/polylactic fibre: this image has been obtained by the author from the Wikimedia website where it was made available by Polimerek (2008). It is stated to be in the public domain. It is included within this chapter on that basis. It is attributed to Polimerek. Lyocell: this image has been obtained by the author from the Wikimedia website where it was made available by FChlo (2024). It is stated to be in the public domain. It is included within this chapter on that basis. It is attributed to FChlo. Rayon: Reproduced from ChemWhat.)

Natural polymers	Structural formula	Raw material	Chemical process	Trade names
Cellulose acetate [113]		Wood, cotton linters	Acetylation	Setilithe Plastiloid Bioceat

Cellulose triacetate [113]	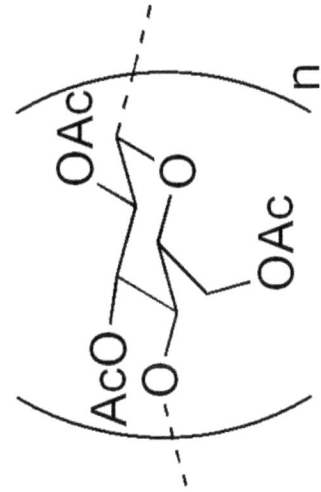 Ac = OOCCH3	Wood, cotton linters	Acetylation	Tricel Arnel
Viscose or rayon [113, 128, 129]	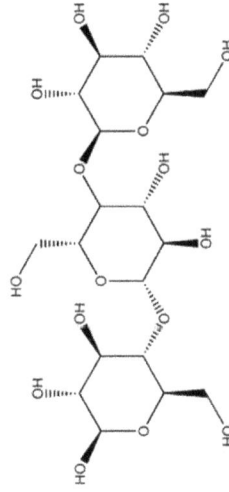	Bamboo, wood	Xanthation	Viscose Rayon

(Continued)

Table 2.10. (*Continued*)

Natural polymers	Structural formula	Raw material	Chemical process	Trade names
Lyocell [113]		Eucalyptus + chitosan (Tencel) Beech wood (Modal) Seaweed + cellulose (Seacell)	Paper dissolved in N-methyl morpholine-N-oxide	Tencel Modal Seacell
Corn fibre (polylactic acid fibre) [113]		Corn	Fermentation and separation	Ingeo Biophyl Lactron Ecodear Sorona

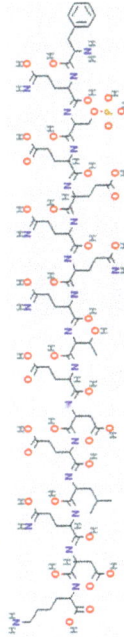

| Soya fibre [130–132] | Soya bean | Denaturation | SoySilk Silkool Winshow |
| Milk fibre (cassein) [124, 129] | Sour milk | Fermentation | Aralac Lanital Lactofil Casolana Tiolan Polana Cargan Merinova Firbolan |

Table 2.11. Protein (keratin-based) fibres from animals. (The schematic diagram of wool fibre structure has been obtained by the author from the Wikimedia website where it was made available by CSIRO (2020) under a CC BY 3.0 licence. It is included within this chapter on that basis. It is attributed to CSIRO.)

Animal fibre (keratin-based)	Origin of fibre	Features	Length of fibre	Diameter of fibre
Alpaca [140]	*Vicugna pacos* (South American camelid)	Dry fibre, minimum lanolin content	8 cm +	15–16 μm
Angora [140, 141]	Angora rabbit	Fur fibre, dry fibre	<20 mm	14–16 μm
Camel [140, 141]	Bactrian camel	Down fibre	2.5–15 cm	10–40 μm
Wool [140]	Sheep	Wool with lanolin	38–375 mm	17–40 μm
Chiengora	Dog		Varies widely	Varies widely
Llama [140]	Llama	Dry fibre, minimum lanolin content	30 cm	20–27 μm
Mohair [140, 141]	Angora goat	Fur fibre, high content of grease	10–30 cm	23–38 μm
Qiviut [142]	Musk oxen	Down fibre	5.2 cm	16–18 μm
Vicuna [140, 143]	Wild version of alpaca	Down fibre	30 mm	12 μm
Cashmere [140]	Cashmere goat	Down fibre	2.5–9 cm	15 μm
Yak [144]	Yak	Down fibre	20–60 mm	15–19 μm
Silk	*Bombyx mori*	Protein fibre	300–900 m	10 μm

Table 2.12. Hydrocarbon based fibres. (Acrylic: this image has been obtained by the author from the Wikimedia website where it was made available by Chem Sim 2001 (2019). It is stated to be in the public domain. It is included within this chapter on that basis. It is attributed to Chem Sim 2001. Poly-paraphenylene/ Kevlar: this image has been obtained by the author from the Wikimedia website where it was made available by Ben Mills and Jynto (2010). It is stated to be in the public domain. It is included within this chapter on that basis. It is attributed to Ben Mills and Jynto. Nylon 6,6: this image has been obtained by the author from the Wikimedia website where it was made available by D.328 (JChemPaint) (2005), under a CC BY-SA 3.0 license. It is included within this chapter on that basis. It is attributed to D.328 (JChemPaint). PET: this image has been obtained by the author from the Wikimedia website where it was made available by Project Osprey (2021) under a CC BY-SA 4.0 license. It is included within this chapter on that basis. It is attributed to Project Osprey.)

Synthetic polymers	Origin of fibre	Trade name
Acrylic	Acrylonitrile monomer	Orlon Acrilan Courtelle Creslan Dynel
Poly-paraphenylene terephthalamide	The reaction of 1,4-phenylene-diamine (*para*-phenylenediamine) with terephthaloyl chloride yielding Kevlar	Kevlar Twaron
Poly (m-phenylenediamine isophthalamide)	www.explainthatstuff.com	Nomex Teijinconex

(Continued)

Table 2.12. (*Continued*)

Synthetic polymers	Origin of fibre	Trade name
Nylon	Nylon 6,6 structure	Nilit Cordura Ultron
Polyethylene terephthalate (PET)		Polyester Terylene Lavsan Dacron Tetoron Diolen Elana Tergal
Spandex	Polyether-polyurea copolymer	Lycra Elastan Elaspan Acepora Creora INVIYA ROICA Dorlastan Linel ESPA

- Protein based sizing agents such as gelatin, glue and albumen.
- Solvent-soluble materials (for example copolymers of methylacrylate, modified polyesters, and maleic acid copolymers).

100% cotton yarn is sized mostly with starch or starch-based sizes since they are economical and sustainable, whereas blends are sized with organic polymers, for example poly(vinyl alcohol) is widely used for cotton/polyester blends [133] or solvent-soluble sizes. The most common desizing agents today are α-amylases, enzymes degrading starch and starch derivatives. Traditionally they are used in warm water at 50 °C–70 °C, see figure 2.64. The fabric is left in the bath for several hours and then washed and rinsed.

Grey fabric ──▶ Desizing ──▶ Washing and Rinsing

Scouring ────────▶ Washing and Rinsing

Bleaching ────────▶ Washing, Rinsing and Neutralisation

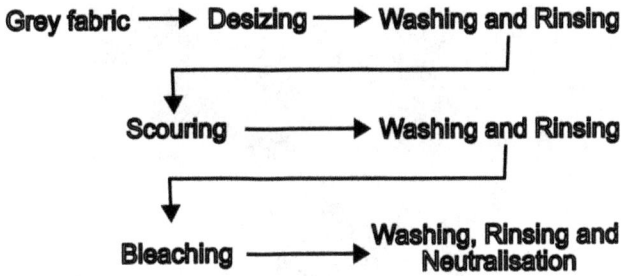

Figure 2.64. Conventional pre-treatment steps for cotton as described in [135].

3. *Scouring*

Scouring is the treatment of fabric with hot alkaline solutions, for example caustic soda (sodium hydroxide) or, when only waxes and fats are present as impurities, trichloroethylene. It removes the impurities from the fabric. Natural cellulosic fabric loses about 5%–10% of its weight in the process and the wettability and absorbance of the fibres are dramatically improved [133]. Scouring can be replaced by mercerizing where the fabric is soaked in a caustic solution and then stretched.

4. *Bleaching*

After washing and rinsing, the fabric is bleached to achieve a neutral background and batch stability for printing. The most important bleaching agent is hydrogen peroxide because it is environmentally safe and versatile. It can be used in hot or cold water, in rapid or long-dwell processes, and batchwise or continuously. The fabric is then washed one last time and dried. It is then ready for printing.

Other cellulosic fibres, including regenerated cellulosic fibres, are treated in similar ways. The water consumption of the pre-treatment process is substantial.

All fabric, except felted fabric, is constructed out of yarns or threads. A yarn consists of fibres which have been spun together to form a long, continuous thread of interlocking fibres. The surface chemistry of those fibres is the determining factor for any kind of fastness of a print on a fabric. Printing on textiles can be understood as a form of localized dying. Different to dying, during printing the dyestuff is applied as a thickened paste. Its viscosity is crucial to the success of the print since it determines the volume transferred to the fabric, how much it penetrates the fabric and how much it spreads or bleeds in the fibre [136]. With the majority of fabric printing done as screen printing, pigment is the most common dyestuff for printing on textiles. The pigment itself does not interact much with the fibre, but it is held in place or glued to the surface of the fibre, which can include internal surfaces as well, by a heat cured binder. There are two big advantages of pigment printing. (i) Since the pigments are defined as insoluble dye particles which do not interact chemically with the fibre, the same pigment can be used for different fabrics. Only the binder sometimes needs a little bit of tweaking.

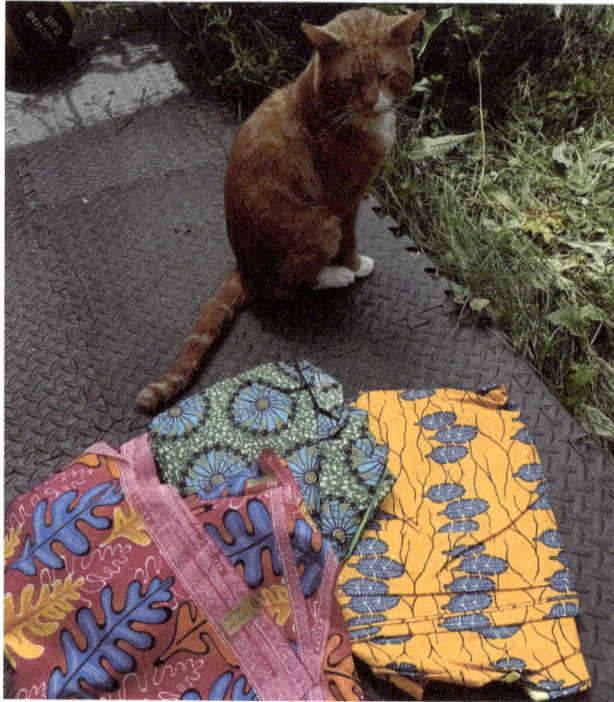

Figure 2.65. Examples of Ankara or Dutch wax print.

(ii) The fabric does not need to be washed after the fixation process [136]. Pigment and dye printing are so-called direct printing methods. Before direct printing is discussed in more detail, we would like to mention two other printing methods which are a combination of printing and dying: discharge printing and resist printing. In discharge printing, the fabric is dyed first. A paste is printed on the fabric which bleaches the dye the fabric was dyed with and leaves a light pattern on the coloured surface. The paste can contain other dyestuff which then immediately stains the bleached areas. In resist printing a pattern is printed on the fabric which prevents the dye from penetrating or the pigment from attaching to the fabric when immersed in the dye bath. A classic example is the Ankara or Dutch wax print, see figure 2.65, where the process is repeated several times. Another printing method favoured for small runs is sublimation printing. Sublimation printing is an indirect printing method. The design is inkjeted onto a transfer film and then attached to a polyester fabric via heat. This method is not suitable for natural fibres.

When printing with dyes, the surface chemistry of the fibre has to be suitable for the interaction with the dye molecules. Fibres can be divided in a hydrophilic and a hydrophobic group, see table 2.13.

Hydrophilic fibres can be printed with water-based pastes, hydrophobic fibres cannot.

Table 2.13. Grouping of textile fibres into hydrophilic and hydrophobic.

	Hydrophilic
Cellulosic fibres	Vegetable fibres: cotton, linen, hemp, jute, nettle, ramie, kenaf, banana sisal, abaca, piña, raffia, coir, kapok, bamboo
	Regenerated cellulose: viscose, Lyocell
Protein fibres	Alpaca, angora, camel, wool, chiengora, llama, mohair, qiviut, vicuna, cashmere, yak, silk, casein fibre, soya fibre
Other fibres	Corn fibre
	Hydrophobic
Synthetic polymer fibres	Non-ionic: secondary cellulose acetate, cellulose triacetate, polyester, polypropylene Cationic: polyamide (nylon)
	Anionic: polyacryonitriles (acrylics)

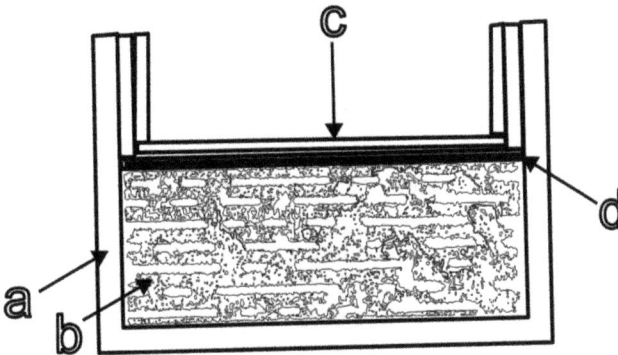

Figure 2.66. The swimming tub or sieve: (a) wooden tub, (b) thickening or old dye paste, (c) drum stretched with a woollen cloth on which the dye paste is applied with a brush, (d) a drum stretched with a waterproof covering separating (c) from (b). (Based on the description in [137].)

2.4.2 Direct print on textiles

2.4.2.1 Block printing

As mentioned before, the oldest method of textile printing is block printing, a technique which still exists as an artisan industry. Block printing is a relief printing method: the ink sits on the highest parts of the printing block. The ink, a paste of either dye or pigment, is applied to the block via the so-called 'swimming tub' or 'sieve' [137], a wooden tub which is constructed in such a way that it allows the block to be inked evenly. The European version is shown in figure 2.66.

An assistant brushes the ink paste on the woollen cloth (c) and the printer inks the block by pressing the block onto (c) (figure 2.67). The block is then positioned on the

Figure 2.67. Hand block printing. (Reproduced from Yoshihiro Kudo on flickr (2015), CC BY-SA 2.0.)

fabric substrate and hit with either the hand or the handle of a mallet to imprint the pattern, see figure 2.67

In the industrial revolution hand block printing was replaced by engraved roller printing, which is an intaglio printing process, i.e. the ink sits in the depressions of the printing plate. The plate is mounted on a roller allowing a much higher throughput (figure 2.68).

Figure 2.68 shows the principles of engraved roller printing and figure 2.69 the set up of a multi-colour machine.

2.4.2.2 Screen printing

Rotary screen printing is the most prominent printing method for high volume textile printing. Screen printing is a stencil method where the image is generated by closing the openings in a mesh. This can be done either by applying a stencil or by curing a photosensitive layer on the mesh. The ink is then forced through the open parts of the mesh.

Figure 2.70 shows an example of the simplest form of screen printing. The frame with the pattern is moved along the fabric and the ink is forced through the mesh with the help of a so-called squeegee. Comparing the size of the block in figure 2.67 and the screen in figure 2.70 it becomes clear that even the intensive hand printing method of screen printing is faster than block printing.

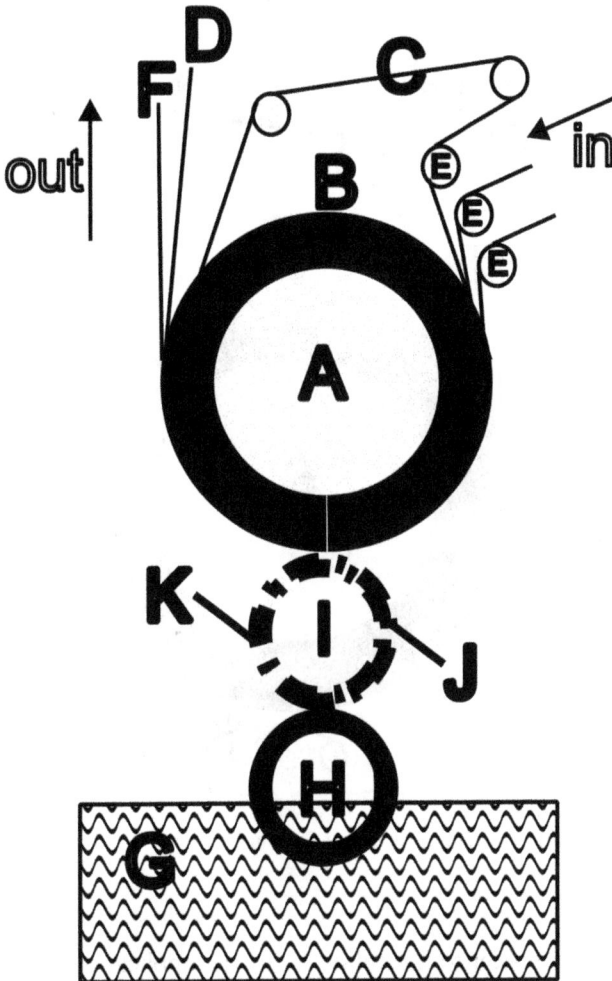

Figure 2.68. (A) Pressure cylinder covered with (B) layers of fabric. (C) Endless blanket. (F) Fabric that is printed with 'back grey' cotton fabric (D) which protects C and B from staining. (E) Tension rollers. (I) Cylinder with pattern. (J) Doctor blade which removes excess ink. (K) Doctor plate which removes lint. (H) Furnishing roller which transfers the ink from trough (G) to (I).

2.4.2.3 Digital printing

Digital printing uses inkjet technology to transfer a pattern directly onto the textile without any plates or screens. As with all digital printing the method is most suitable for small runs and customization, because there is no need for any printing plates or screens. The size of the print can be changed as long as it does not exceed the physical constraints of the printer itself. The inks need to be suitable for the printing method.

For inkjet the pigments must be relatively small and suspended well in the binder. The inks should neither clog the nozzle nor dry in the nozzle. The drop formation should be reliable. Pigmented inks are held by the binder on the fibre and used on natural fibres. A rotary heat calender fixes the print onto the fabric.

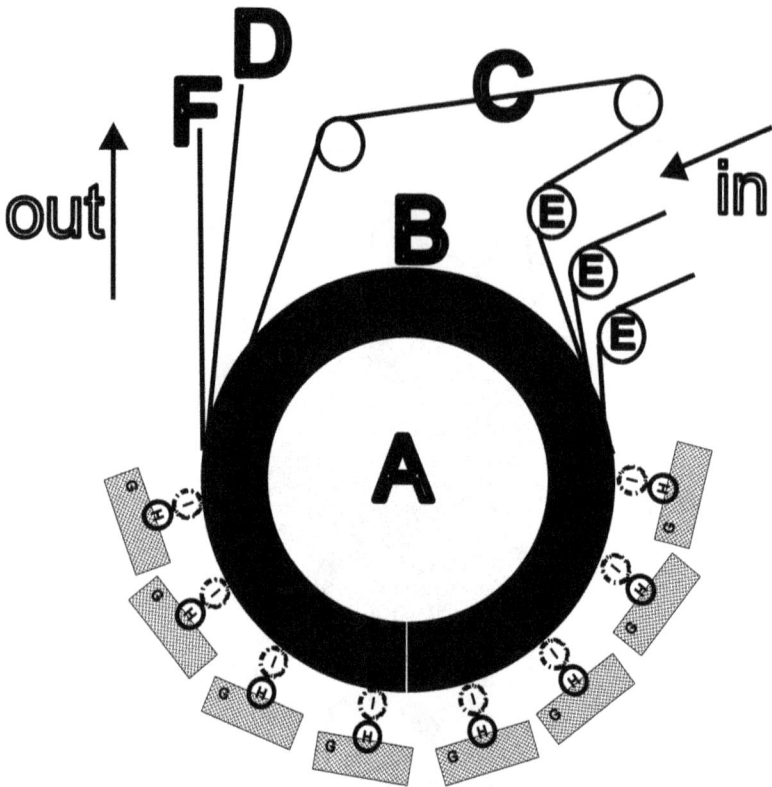

Figure 2.69. In a multi-colour roller, machine trough G, furnishing roller H and printing roller I are repeated for each colour.

Dye based inks are easier to formulate for ink jet applications and need the usual post-processing. Direct-disperse inks are used for polyester or polyester blends. The print has to be fixed with a heat treatment.

Acid dyes are used on nylon and silk. The printed fabric must be steamed, washed and heat treated to make the print permanent.

Reactive inks: are used for protein and cellulose fibers. The fabric is pretreated before the inks are jetted onto it. To make the ink react with the fibers, the print must be steamed. The fabric is then washed to remove the pretreatment and surplus ink.

A variation of digital printing is dye sublimation printing where the image is digitally printed on a carrier substrate and is then transferred by heat to the fabric. Sublimation printing works only on polyester or polyester containing fabrics. As with direct digital print the method is most useful for small runs and customization.

Printing on textiles is not only governed by the application but also by the substrate. Textiles can have many different surfaces and there is neither an ink nor a printing method which covers all of them. Printing on textiles is not only decorative but can also be functional. Electronic textiles have attracted a lot of scientific and commercial interest as

Figure 2.70. Hand screen printing. (US National Archives and Records Administration (210-G-E700). Image stated to be in the public domain.)

they could supply the next generation of sensors and energy generators [138]. In addition to weaving and embroidery, printing is a commercially attractive method to equip garments with additional functionality because it is a fast and cost effective process.

References

[1] Kang T 2007 Role of external fibrillation in pulp and paper properties *PhD thesis* Helsinki University of Technology http://lib.tkk.fi/Diss/2007/isbn9789512289172/

[2] Motamedian H R, Halilovic A E and Kulachenko A 2019 Mechanisms of strength and stiffness improvement of paper after PFI refining with a focus on the effect of fines *Cellulose* **26** 4099–124

[3] Ek M, Gellerstedt G and Henriksson G 2009 15.10 How many times can a fibre be reused? *Pulp and Paper Chemistry and Technology—Pulping Chemistry and Technology* **vol 2** (Berlin: De Gruyter)

[4] 1996 *Handbook of Pulping and Papermaking* 2nd edn ed C J Biermann (Amsterdam: Elsevier) p 234

[5] Smook G A 2016 *Handbook for Pulp and Paper Technologists* 4th edn (Peachtree Corners, GA: Tappi) p 438

[6] Steven Mann Paper Classroom https://www.youtube.com/user/PaperClassroom (YouTube channel)

[7] Smook G A 2016 17.1.2 Zones of evaporation *Handbook for Pulp and Paper Technologists* 4th edn (Peachtree Corners, GA: Technical Association of the Pulp and Paper Industry (TAPPI)) p 273

[8] Smook G A 2016 17.1.1 Criteria of performance *Handbook for Pulp and Paper Technologists* 4th edn (Technical Association of the Pulp and Paper Industry (TAPPI)) p 272

[9] Browne T C and Crotogino R H 2018 Future directions in calendering research *Trans. of the 12th Fund. Res. Symp. The Science of Papermaking (Oxford, 2001)* C F Baker (FRC: Manchester) pp 1001–36

[10] Roberts J C 1996 *The Chemistry of Paper* (Cambridge: Royal Society of Chemistry)

[11] Papyrus making 101 *Papyrus Collection, University of Michigan* https://www.lib.umich.edu/papyrus_making/index.html

[12] Statista Research Department 2020 Global paper industry—statistics and facts *Report* Statista https://www.statista.com/topics/1701/paper-industry/

[13] P G Paper 2018 The global paper market—current review *Report* https://www.pgpaper.com/wp-content/uploads/2018/07/Final-The-Global-Paper-Industry-Today-2018.pdf

[14] Berg P and Lingqvist O 2019 Pulp, paper and packaging in the next decade: transformational change *Paper, Forest Products and Packaging Article* McKinsey https://www.mckinsey.com/industries/paper-forest-products-and-packaging/our-insights/pulp-paper-and-packaging-in-the-next-decade-transformational-change#

[15] Liu Z, Wang H and Hui L 2018 Pulping and papermaking of non-wood fibers *Pulp and Paper Processing* ed S N Kazi (London: IntechOpen)

[16] Ek M, Gellerstedt G and Henriksson G 2009 *Pulp and Paper Chemistry and Technology* (Berlin: de Gruyter)

[17] Ali Z M and Gibson L J 2013 The structure and mechanics of nanofibrillar cellulose foams *Soft Matter* **9** 1580–8

[18] Ghosh A K 2011 Fundamentals of paper drying: theory and application from industrial perspective *Evaporation, Condensation and Heat Transfer* ed A Ahsan (London: IntechOpen)

[19] Brandberg A, Motamedian H R, Kulachenko A and Hirn U 2020 The role of the fiber and the bond in the hygroexpansion and curl of thin freely dried paper sheets *Int. J. Solids Struct.* **193–194** 302–13

[20] Page D H and Tydeman P A 1962 A new theory of the shrinkage, structure and properties of paper *Trans. of the IInd Fund. Res. Symp. The Formation and Structure of Paper* vol 1 *(Oxford, 1961)* F Bolam (Manchester: FRC) pp 397–425

[21] Nanko H 1995 Mechanisms of hygroexpansion of paper *International Paper Physics Conf. (Niagara-on-the-Lake, ON, Canada)* pp 159–71

[22] Nanko H 1995 Mechanisms of paper shrinkage during drying *International Paper Physics Conf. (Niagara-on-the-Lake, ON, Canada)* pp 103–13

[23] Seth R S 2005 Understanding sheet extensibility *Pulp Paper Canada* **106** 33–40

[24] Linhart F 1995 The practical application of wet-strength resins *Applications of Wet-End Paper Chemistry* ed C O Au and I Thorn (Dodrecht: Springer)

[25] Cantero D 2014 Intensification of cellulose hydrolysis process by supercritical water. Obtaining of added value products *PhD in Thermodynamics Engineering of Fluids* Universidad de Valladolid, Spain

[26] Hu G, Fu S and Liu H 2013 Hemicellulose in pulp affects paper properties and printability *Appita J.* **66** 139–44

[27] Sitch D A and Marshall H B 1950 The effect of hemicelluloses on the papermaking properties of white birch *Can. J. Res.* **28f** 376–89

[28] Mobarak F, El-Ashmawy A E and Fahmy Y 1973 Hemicelluloses as additive in paper-making. II. The role of added hemicellulose, and hemicellulose in situ on paper properties *Cellul. Chem. Technol.* **7** 325–35

[29] Bhaduri S K, Ghosh I N and Deb Sarkar N L 1995 Ramie hemicellulose as beater additive in paper making from jute-stick kraft pulp *Ind. Crops Prod.* **4** 79–84

[30] Korbag I and Mohamed S 2014 Extraction of lignin from paper industry waste *Int. J. Appl. Eng. Res.* **9** 19421–8

[31] Hubbe M, Alén R, Paleologou M, Kannangara M and Kihlman J 2019 Lignin recovery from spent alkaline pulping liquors using acidification, membrane separation, and related processing steps: a review *BioResources* **14** 2300–51

[32] Humpert D, Ebrahimi M and Czermak P 2016 Membrane technology for the recovery of lignin: a review *Membranes* **6** 42

[33] Kevlich N S, Shofner M L and Nair S 2017 Membranes for kraft black liquor concentration and chemical recovery: current progress, challenges, and opportunities *Sep. Sci. Technol.* **52** 1070–94

[34] Zhao W *et al* 2016 From lignin association to nano-/micro-particle preparation: extracting higher value of lignin *Green Chem.* **18** 5693–700

[35] Upton B M and Kasko A M 2016 Strategies for the conversion of lignin to high-value polymeric materials: review and perspective *Chem. Rev.* **116** 2275–306

[36] Gess J M 1989 Rosin sizing of papermaking fibres *TAPPI J.* **72** 77–9

[37] Hubbe M A 2007 Paper's resistance to wetting—a review of internal sizing chemicals and their effects *Bioresources* **2** 106–45

[38] Samyn P 2013 Wetting and hydrophobic modification of cellulose surfaces for paper applications *J. Mater. Sci.* **48** 6455–98

[39] Tester R F, Karkalas J and Qi X 2004 Starch ̶composition, fine structure and architecture *J. Cereal Sci.* **39** 151–65

[40] Murray-Curvex printing machine picking the colour up *Transferware Collectors Club* https://www.transferwarecollectorsclub.org/annex/image-gallery/processes/processes-printing/97-p-trc-611/ (Accessed: 20 September 2023)

[41] Cadmium pigments saved *Cranfield* https://www.cranfield-colours.co.uk/2016/03/24/cadmium-pigments-saved/ (Accessed: 20 September 2023)

[42] Digital Ceramics https://www.digitalceramics.com (Accessed: 20 September 2023)

[43] Mossman S T I 2008 *Fantastic Plastic: Product Design + Consumer Culture* (London: Black Dog)

[44] Wang B, Yang W, McKittrick J and Meyers M A 2016 Keratin: structure, mechanical properties, occurrence in biological organisms, and efforts at bioinspiration *Prog. Mater. Sci.* **76** 229–318

[45] The Copeland Period, 1901–1966 *Spode Museum* https://spodemuseumtrust.org/history/timeline/the-copeland-period-part-2-1901-1966/ (Accessed: 20 September 2023)

[46] Tamburini D, Dyer J and Bonaduce I 2017 The characterisation of shellac resin by flow injection and liquid chromatography coupled with electrospray ionisation and mass spectrometry *Sci. Rep.* **7** 14784

[47] Specification of shellac grades *M/S. D. Manoharlal (Shellac) Pvt. Ltd* https://www.dmshellac.com/specifications/ (Accessed: 27 February 2023)

[48] Rehding A 2006 On the record *Camb. Opera J.* **18** 59–82

[49] Brydson J A 1988 *Rubbery Materials and their Compounds* (London: Elsevier)

[50] Carrillo J 2002 The "Historia General y Natural de las Indias" by Gonzalo Fernández de Oviedo *Huntington Lib. Quart.* **65** 321–44

[51] 2023 Rubber *The Columbia Electronic Encyclopedia* 6th edn (New York: Columbia University Press)

[52] Macintosh C 1823 *UK Patent* 4804

[53] Hancock T 1857 *Personal Narrative of the Origin and Progress of the Caoutchouc or India-rubber Manufacture in England* (London: Longman, Brown, Green, Longmans, and Roberts)

[54] Goodyear C 1844 Improvement in India-rubber fabrics *US Patent Specification* US3462A

[55] Vitreous enamel *A J Wells* https://www.ajwells.com/services/vitreous-enamel/ (Accessed: 20 September 2023)

[56] Virdee S S and Thomas M B M 2017 A practitioner's guide to gutta-percha removal during endodontic retreatment *Br. Dent. J.* **222** 251–7

[57] Rasmussen S C 2021 From Parkesine to celluloid: the birth of organic plastics *Angew. Chem. Int. Edn Engl.* **60** 8012–6

[58] Parkes A 1865 Improvements in the manufacture of Parkesine or compound of pyroxyline, and also solutions of pyroxyline also known as collodion *GB Patent Specification* 1313

[59] Hyatt J, John W and Hyatt I S 1870 Improvement in treating and molding pyroxyline *US Patent Specification* US133229A

[60] McCord C P 1964 Celluloid. The first american plastic—the world's first commercially successful plastic *J. Occup. Med.* **6** 452–7

[61] Jenkins R V 1975 Technology and the market: George Eastman and the origins of mass amateur photography *Technol. Cult.* **16** 1–19

[62] Goodwin H 1898 Photographic pellicle and process of producing same *US Patent Specification* US610861A

[63] Carter E A, Swarbrick B, Harrison T M and Ronai L 2020 Rapid identification of cellulose nitrate and cellulose acetate film in historic photograph collections *Herit. Sci.* **8** 51

[64] Fischer S, Thümmler K, Volkert B, Hettrich K, Schmidt I and Fischer K 2008 Properties and applications of cellulose acetate *Macromol. Symp.* **262** 89–96

[65] Türk O 2014 *Stoffliche Nutzung nachwachsender Rohstoffe: Grundlagen—Werkstoffe—Anwendungen* 1st edn (Wiesbaden: Springer Vieweg) (in German)

[66] Baekeland L 1907 Method of making insoluble products of phenol and formaldehyde, *US Patent Specification* 942699

[67] Conradie N, Meyer S, Rossow J and Van Zyl H Structure *Bakelite* (University of Stellenbosch) https://bakelitegroup62.wordpress.com/category/structure/ (Accessed: 7 August 2023)

[68] American Chemical Society 1993 The Bakelizer *Commemorative Booklet* https://www.acs.org/education/whatischemistry/landmarks/bakelite.html

[69] Ciardelli F, Bertoldo M, Bronco S and Passaglia E 2019 *Polymers from Fossil and Renewable Resources: Scientific and Technological Comparison of Plastic Properties* 1st edn (Cham: Springer International)

[70] Tripathi D 2002 *Practical Guide to Polypropylene (Rapra Practical Guide Series)* (Akron, OH: Smithers Rapra Technology)

[71] Hindle C Polypropylene (PP) *British Plastics Federation* https://www.bpf.co.uk/plastipedia/polymers/PP.aspx (Accessed: 9 August 2023)

[72] Federation B P 2023 Polyvinyl chloride (PVC) *British Plastics Federation* https://www.bpf.co.uk//plastipedia/polymers/PVC.aspx (Accessed: 10 August 2023)

[73] ChemViews 2023 *100th Anniversary of the First PVC Patent* (Weinheim: Wiley) https://www.chemistryviews.org/details/ezine/4899111/100_Anniversary_of_the_First_PVC_Patent/ (Accessed: 10 August 2023)

[74] Simon E 1839 Uber den flussigen Storax (Styrax liquidus) *Ann. Chem.* **31** 265–77

[75] Demirors M 2000 Styrene polymers and copolymers *Applied Polymer Science: 21st Century* ed C D Craver and C E Carraher (Oxford: Pergamon) pp 93–106

[76] McIntire O R 1948 Manufacture of cellular thermoplastic products *US Patent Specification* US2450436

[77] American Chemical Council 2023 Polystyrene *Chemical Safety Facts* https://www.chemicalsafetyfacts.org/chemicals/polystyrene/ (Accessed: 14 August 2023)

[78] Plastics Europe 2023 Plastics—The facts 2022 *Report* https://plasticseurope.org/knowledge-hub/plastics-the-facts-2022-2/ (Accessed: 18 August 2023)

[79] Whinfield J R and Dickson J T 1949 Polymeric linear terephthalic esters *US Patent Specification* 2465319A

[80] De Vos L, Van de Voorde B, Van Daele L, Dubruel P and Van Vlierberghe S 2021 Poly(alkylene terephthalate)s: from current developments in synthetic strategies towards applications *Eur. Polym. J.* **161** 110840

[81] Wyeth N and Roseveare R 1970 Biaxially oriented poly(ethylene terephthalate) bottle *US Patent Specification* 3733309A

[82] The Editors 2023 Polyethylene terephthalate *Encyclopaedia Britannica* https://www.britannica.com/science/polyethylene-terephthalate (Accessed: 23 August 2024)

[83] Farben I G 1937 Verfahren zur Herstellung von Polyurethanen bezw. Polyharnstoffen *German Patent Specification* DE728981

[84] Bayer O 1947 Das Di-Isocyanat-Polyadditionsverfahren (Poyurethane) *Angew. Chem.* **59** 257–88

[85] What is polyurethane? *polyurethanes.org* https://www.polyurethanes.org/what-is-it/ (Accessed: 18 July 2023)

[86] American Chemical Society 2023 Wallace Carothers and the Development of Nylon *Commemorative Booklet* https://www.acs.org/education/whatischemistry/landmarks/carotherspolymers.html#walla ce-carothers-joins-dupont (Accessed: 29 August 2023)

[87] Transparenz und funktionalität für technik und designplexiglas® *Evonik* https://history.evonik.com/de/erfindungen/plexiglas (Accessed: 30 August 2023)

[88] Bottenbruch L, Krimm H and Schnell H 1953 Verfahren zur Herstellung thermoplastischer Kunststoffe *German Patent Specification* DE971777C

[89] Categorizing surface energy *3M* https://www.3m.com/3M/en_US/bonding-and-assembly-us/resources/science-of-adhesion/categorizing-surface-energy/# (Accessed: 4 September 2023)

[90] Hild F Surface energy of plastics *TriStar* https://www.tstar.com/blog/bid/33845/Surface-Energy-of-Plastics (Accessed: 4 September 2023)

[91] Ügdüler S *et al* 2023 Understanding the complexity of deinking plastic waste: an assessment of the efficiency of different treatments to remove ink resins from printed plastic film *J. Hazard. Mater.* **452** 131239

[92] Robinson S 1969 *A History of Printed Textiles* (London: Studio Vista)

[93] Edwards E 2019 *Block Printed Textiles of India: Imprints of Culture* (New Delhi: Niyogi)

[94] Gleba M and Mannering U 2012 *Textiles and Textile Production in Europe from Prehistory to AD 400* Ancient Textiles Series vol 11 (Oxford: Oxbow)

[95] Vanden Berghe I, Gleba M and Mannering U 2010 Dyes: to be or not to be? Investigation of dyeing in Early Iron Age Danish bog textiles *North European Symposium for Archaeological Textiles X*

[96] Vanden Berghe I, Gleba M and Mannering U 2009 Towards the identification of dyestuffs in Early Iron Age Scandinavian peat bog textiles *J. Archaeol. Sci.* **36** 1901–21

[97] Mannering U 2017 *Iconic Costumes: Scandinavian Late Iron Age Costume Iconography* Ancient Textiles Series vol 25 (Oxford/Havertown, PA: Oxbow)

[98] Gleba M 2021 Personal communications

[99] O'Connell E R 2008 Representation and self-presentation in late antique Egypt: 'Coptic' textiles in the British Museum *Textiles as Cultural Expressions: Proceedings of the 11th Biennial Symp. of the Textile Society of America* vol 121 (Honolulu, HI, 24–27 September)

[100] Nelson C N and Johnson R F 1988 Attribute characterization schemes for ancient textiles *Ars Textrina* **9** 11–41

[101] Forrer R 1895 *Mein Besuch in El-Achmim: Reisebriefe aus Aegypten* (Strassburg: Schlesier) (in German)

[102] Khurdadhbib I 885 *Kitab al-Masalik wa al-Mamalik (The Book of Roads and Provinces)* https://archive.org/details/journalasiatique566sociuoft/page/4/mode/2up?view=theater

[103] Rukua Z 1225 *Zhu Fan Zhi (Records of Foreign People)* https://storymaps.arcgis.com/stories/39bce63e4e0642d3abce6c24db470760

[104] Polo M and Rustichello of Pisa 2014 *The Travels of Marco Polo - Complete* N.p. (CreateSpace Independent Publishing Platform)

[105] Jonathan P E 2012 Making an imperial compromise: the Calico Acts, the Atlantic colonies, and the structure of the British Empire *William Mary Quart.* **69** 731–62

[106] Gril-Mariotte A 2015 *Les Toiles de Jouy* (Rennes: Presses Universitaires de Rennes)

[107] Hooker M B 1969 The East India Company and the Crown 1773–1858 *Malaya Law Rev.* **11** 1–37

[108] Wardle Y 1886 *Colonial and Indian Exhibition, 1886. Empire of India. Special Catalogue of Exhibits by the Government of India and Private Exhibitors. Royal Commission and Government of India Silk Culture Court. Descriptive catalogue* (London: Clowes) https://wellcomecollection.org/works/fv64qqsn

[109] Ruskin J 1903–1912 The stones of Venice *The Complete Works of John Ruskin* **vol 10** ed E T Cook and A Wedderburn (London: George Allen and Unwin) p 193

[110] Day L F 1899 The art of William Morris *The Easter Art Annual* 1–32

[111] Our heritage *Liberty Fabrics* https://www.libertyfabric.com/our-heritage/

[112] Grand View Research 2018 Printed textile market size, share and trends analysis report by ink type, by product, by technology, by application (fashion, household, technical textiles), by region, and segment forecasts, 2019–2025 *Report* GVR-3-68038-995-1 https://www.grandviewresearch.com/industry-analysis/printed-textile-market (Accessed: 13 June 2021)

[113] Hallett C and Johnston A 2020 *Fabric for Fashion, The Complete Guide* (London: Laurence King)

[114] Biagiotti J, Puglia D and Kenny J M 2004 A review on natural fibre-based composites—Part I *J. Nat. Fibers* **1** 37–68

[115] Shahzad A 2013 A study in physical and mechanical properties of hemp fibres *Adv. Mater. Sci. Eng.* **2013** 325085

[116] Das S, Singha A K, Chaudhuri A and Ganguly P K 2019 Lengthwise jute fibre properties variation and its effect on jute–polyester composite *J. Text. Inst.* **110** 1695–702

[117] Agus Suryawan I G P, Suardana N P G, Suprapta Winaya I N, Budiarsa Suyasa I W and Tirta Nindhia T G 2017 Study of stinging nettle (*Urtica dioica* L.) Fibers reinforced green composite materials: a review *IOP Conf. Ser.: Mater. Sci. Eng.* **201** 012001

[118] Ji X l, Wu S j and Yu C w 2012 Analysis of ramie fiber length changes during the stretch-breaking process *J. Text. Inst.* **103** 99–105

[119] Discover Natural Fibres Initiative https://dnfi.org/ (Accessed: 11 October 2020)

[120] Textile School https://www.textileschool.com/ (Accessed: 11 November 2020)

[121] Simbaña E A, Ordóñez P E, Ordóñez Y F, Guerrero V H, Mera M C and Carvajal E A 2020 Abaca: cultivation, obtaining fibre and potential uses *Handbook of Natural Fibres* 2nd edn ed R M Kozłowski and M Mackiewicz-Talarczyk (Cambridge: Woodhead) ch 6 pp 197–218

[122] Fibre2Fashion https://www.fibre2fashion.com/ (Accessed: 11 November 2020)

[123] Fadele O, Oguocha I N A, Odeshi A, Soleimani M and Karunakaran C 2018 Characterization of raffia palm fiber for use in polymer composites *J. Wood Sci.* **64** 650–63

[124] CAMEO: Conservation and Art Materials Encyclopedia Online (Museum of Fine Arts, Boston, MA) http://cameo.mfa.org/wiki/Main_Page (Accessed: 16 April 2020)

[125] Mishra L and Basu G 2020 Coconut fibre: its structure, properties and applications *Handbook of Natural Fibres* 2nd edn ed R M Kozłowski and M Mackiewicz-Talarczyk (London: Woodhead) ch 8 pp 231–55

[126] Wang G and Chen F 2017 Development of bamboo fiber-based composites *Advanced High Strength Natural Fibre Composites in Construction* ed M Fan and F Fu (London: Woodhead) ch 10 pp 235–55

[127] Clarke W 1974 *An Introduction to Textile Printing* 4th edn (London: Newnes-Butterworths)

[128] Collier A M 1983 *A Handbook of Textiles* (Leeds: Arnold-Wheaton)

[129] PubChem National Institute of Health (NIH) https://pubchem.ncbi.nlm.nih.gov/ (Accessed: 20 November 2020)

[130] Li G 2007 Phytoprotein synthetic fibre and method of manufacture thereof *US Patent Specification* 10/883,607

[131] Rijavec T and Zupin Z 2011 Soybean protein fibres (SPF) *Recent Trends for Enhancing the Diversity and Quality of Soybean Products* (London: IntechOpen) ch 23 p 501

[132] 1995 *Cellulosic Dyeing* ed J Shore (Bradford: Society of Dyers and Colourists)

[133] Textile Exchange 2021 Preferred fiber and materials *Market Report 2020* https://textileexchange.org/app/uploads/2021/04/Textile-Exchange_Preferred-Fiber-Material-Market-Report_2020.pdf

[134] Harane R S and Adivarekar R V 2016 Sustainable processes for pre-treatment of cotton fabric *Text. Cloth. Sustain.* **2** 2

[135] Broadbent A D 2001 *Basic Principles of Textile Coloration* (Bradford: Society of Dyers and Colourists)

[136] Storey J 1992 *The Thames and Hudson Manual of Textile Printing* rev. edn (London: Thames and Hudson)

[137] Islam M R, Afroj S, Novoselov K S and Karim N 2022 Smart electronic textile-based wearable supercapacitors *Adv. Sci.* **9** 2203856

[138] Boustead I 2005 Polyamide 66 (nylon 66). Eco-profiles of the European Plastics Industry *Report* Plastics Europe http://www.inference.org.uk/sustainable/LCA/elcd/external_docs/n66_311147f8-fabd-11da-974d-0800200c9a66.pdf

[139] Zhang L and Zeng M 2008 Proteins as sources of materials *Monomers, Polymers and Composites from Renewable Resources* (Amsterdam: Elsevier) ch 23 pp 479–93

[140] Cook G J 2001 *Handbook of Textile Fibres* (Oxford: Woodhead)
Natural Fibres https://web.archive.org/web/20171117095210/http://www.naturalfibres2009.org/en/fibres/index.html (Accessed: 20 February 2020)

[141] Rowell J E, Lupton C J, Robertson M A, Pfeiffer F A, Nagy J A and White R G 2001 Fiber characteristics of qiviut and guard hair from wild muskoxen (*Ovibos moschatus*) *J. Anim. Sci.* **79** 1670–4

[142] Vicuna *Dormeuil* https://dormeuil.com/pages/expertise (Accessed: 20 November 2020)

[143] Li W, Liu X, Liu C, Su X, Xie C and Wei Q 2016 Preparation and characterisation of high count yak wool yarns spun by complete compacting spinning and fabrics knitted from them *Fibres Text. East. Eur.* **24** 30–5

[144] Barbosa J, Conway B and Merchant H 2017 Going natural: using polymers from nature for gastroresistant applications *Br. J. Pharm.* **2**

Chapter 3

The physics and chemistry of printing

3.1 The physics of the transfer of ink to the substrate

Printing is an exercise in sticking and unsticking. In traditional printing methods, the image is transferred from a plate to a substrate and the ink is stuck and unstuck several times, in the simplest case, at least three times: the ink must adhere to the applying tool, then it is transferred to the plate and finally from the plate to the substrate. The adhesion forces in the process always have to be in favour of the next step, otherwise the ink will not transfer, and the image content is not complete when it arrives at its final destination. Non-impact methods, i.e. printing methods without a printing plate or master, have fewer steps, but the ink still has to detach from the image carrier or nozzle and attach to the final substrate.

3.1.1 Adhesion

Do we ever truly touch anything and what are sticky fingers?

We, and all matter, are made of atoms. These are the smallest stable units of material and they consist of mostly nothing. The simplest model of an atom is the Rutherford–Bohr model based on experimental work done by Ernest Rutherford published in 1911 [1] and theoretical work done by Niels Bohr published in 1913 [2]. In this model the atom consists of a nucleus, where protons and neutrons are densely packed, which is orbited by electrons. Since protons and neutrons are in the nucleus, they are called nucleons. They are responsible for almost all the mass of the atom. In addition to adding part of the mass, the protons give the nucleus a positive charge, that is where their name comes from. The orbiting electrons have a negative charge and the interaction between the electrons and protons leads to a neutral charge for the atom. The nucleus is extraordinarily small compared to the overall size of the atom. On average the size of an atom can be detected as 10^{-10} m. The nucleus is between 2×10^{-15} and 12×10^{-15} m [3]. Imagine you are the nucleus and are standing at the harbour in Holyhead on the Welsh island of Anglesey. Your atom

doi:10.1088/978-0-7503-2568-4ch3

3-1

would reach as far as Dublin and there would be very little between you and Dublin (figure 3.1). Atoms are mostly nothing with a very heavy centre.

Adhesion happens in Dublin, not in Holyhead, but how close can we get?

A simple mathematical model for the interaction between two neutral atoms is the Lennard-Jones model. In [4] J E Jones (later, after marrying Kathleen Lennard, Lennard-Jones) developed a mathematical formula which describes what happens when two neutral atoms interact within a gas, i.e. when they can move freely and are not locked into a crystal lattice or a molecular chain. The formula for the Lennard-Jones potential is

$$V(r) = 4\varepsilon\left[\left(\frac{\sigma}{r}\right)^{12} - \left(\frac{\sigma}{r}\right)^{6}\right],$$

where r is the distance between the two atoms, ε is the depth or minimum value of the potential, and σ is the distance where the potential between the two atoms is zero, i.e. they are just touching. For two identical atoms σ would be two times the radius.

A potential or potential energy describes the work which must be done when the two atoms change distance. The force needed can be calculated from the gradient of the potential. The steeper the gradient the more force is needed. By convention a negative potential describes attraction, a positive one repulsion.

Figure 3.1. If the reader is the nucleus and is standing at the harbour in Holyhead, the atom would extend to the circle shown in the image.

Lennard-Jones Potential

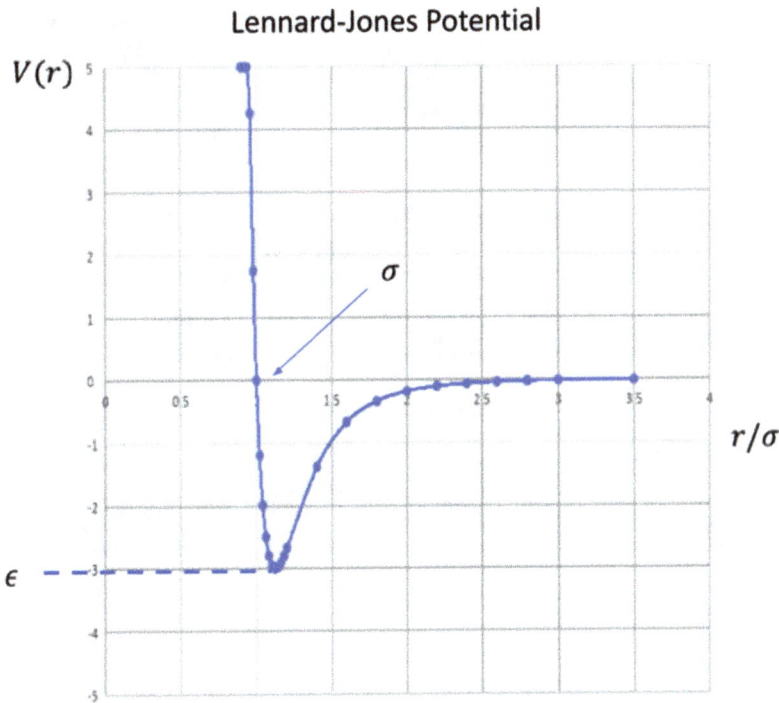

Figure 3.2. Lennard-Jones potential; ε is the depth or minimum value of the potential and σ is the distance where the potential between the two atoms is zero.

In figure 3.2 we can see that for two identical atoms beyond $3r/\sigma$, i.e. six times the radius of the atom, the potential is still negative, but the gradient is rather flat. Moving the atoms further apart does not require much force. When the atoms come closer than five times their radius, the gradient increases. The atoms are attracted to each other since the potential is negative and keeping them apart requires force. When the minimum value of the potential is reached, the gradient is again 0. No force is required to keep them there and they will stay at this distance unless energy is pumped into the system to either push them further together or to push them apart. Pushing them closer than σ is very difficult. The atoms repel each other, and the gradient is very steep. A lot of force is needed. It can be achieved, but it does not happen during printing. The energy required to bring two atoms closer together than σ can only be achieved in big accelerators such as CERN.

We know now that all atoms stick when they are close enough to sit in the Lennard-Jones potential well. This atomic or molecular attraction happens when bodies make close contact, so close that the gaps are near molecular dimensions [5]. The range of this force is so short that even micrometre gaps prevent sticking. To allow molecular adhesion, surfaces have to be very flat and very clean. A situation not often encountered in printing.

So how does the ink stick? Let's consider the process of hand offset lithography. In hand offset lithography ink is transferred five times (see figure 3.3):

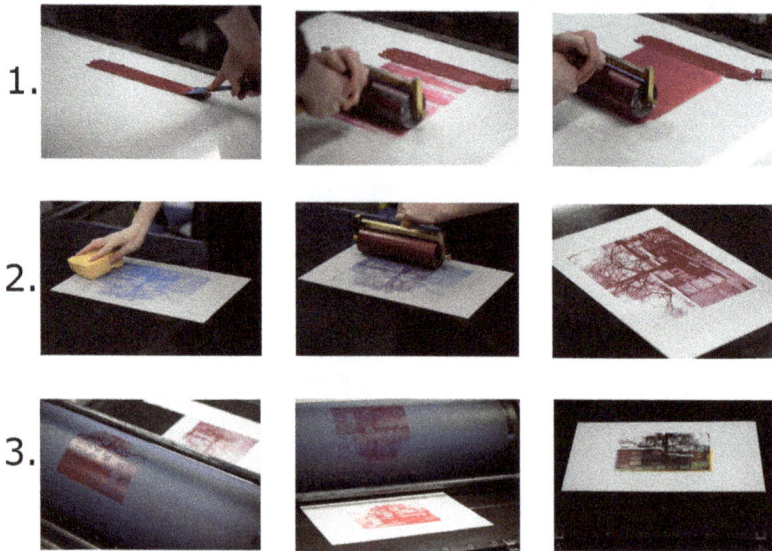

Figure 3.3. Hand offset printing: 1. Spreading the ink onto a glass surface. 2. Inking the plate. 3. Ink transfer from the plate to the blanket and from the blanket to the paper. (Images taken by and permission obtained from Frank Menger.)

1. The ink is lifted with a spatula out of the tin and a line of ink is spread on a glass plate.
2. With a rubber roller it is then smoothed into a velvety layer. Different strengths of adhesion can be felt during the process. When the roller is only partly wetted by ink, it will stick firmly to the glass surface at its not-yet-inked parts. When the ink is spread evenly on the glass surface, the roller will roll smoothly and with little resistance.
3. The printing plate is first wiped with a wet sponge. Lithography works on the principal that the oily ink will stick to hydrophobic parts of the plate and not to the hydrophilic parts. The wet plate is then rolled with the inked roller and the process is repeated until all ink holding parts of the image are covered with ink.
4. The image is then transferred to the so-called blanket, a big rubber sheet on a roller across the whole width of the printing press.
5. Finally, the image is transferred to the image carrier, here paper.

The ink for hand offset lithography is a paste made of linseed oil and pigment. The pigment load is between 20 and 40 wt%. The pigment particles are not stabilized, and phase separation will take place eventually, but slowly. Depending on the formulation, it will take weeks or even years. A paste is an intermediate between a sol and a gel (figure 3.4).

In a gel the adhesion between the particles is strong and the particles in a gel form a network. In a sol the adhesion force between the particles should be zero. In a paste

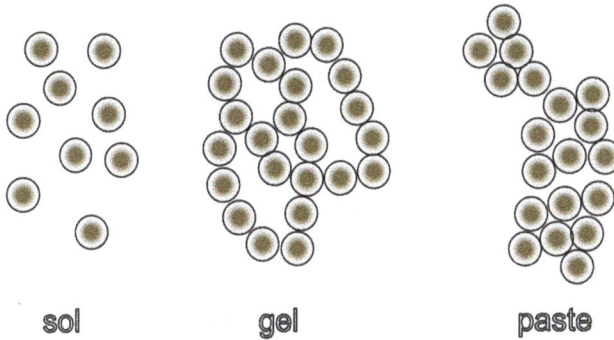

Figure 3.4. Configurations of particles in sol, gel and paste. In a sol the particles are not touching. In a gel they form networks and in a paste they form aggregates [5].

the adhesion between particles is of medium strength, strong enough to form aggregates, i.e. localized gels, but weak enough to allow these aggregates to break apart under shear.

Let us describe the very first step of the printing process—taking the ink out of the pot—because it will help us to understand all the forces involved later in the process. A spatula is pushed into the ink and then the ink is then scooped out of the container. The pot containing the ink is held down during this process. Most inks for hand printing are so viscous that the container will lift when it is not held down. A typical container holds about 500 g of ink. The gravitational force of 500 g ink is

$$F_g = m\,g = 0.5 \times 9.807\,\text{N} = 4.9\,\text{N},$$

where m in kg is the mass of the ink and g is the free fall acceleration with an average value of $9.807\,\text{m s}^{-2}$.

The cohesion within the ink is also at least 4.9 N over the area of the spatula which is pushed into the ink and then lifted out. The area of the spatula in contact of the ink is roughly $0.001\,\text{m}^2$. We need to overcome a minimum of $4900\,\text{N m}^{-2}$ to break the cohesion between the ink particles and to split the ink during the different printing steps.

When the ink is taken out of the container, it is rather stiff and has to be softened by working it with a spatula on a glass plate. This breaks the networks in the ink apart and, since the viscosity of the ink is rather high, it will take a while until they reform. A line of ink is then put down on the glass plate and spread with a polymer roller until an even and velvety ink layer has formed on the roller and glass plate (the sound the roller makes when spreading the ink on the glass plate will tell the printmaker when the right amount of ink is spread for a successful print). When the roller is not completely inked, it sticks to the glass with more adhesion than when it is completely inked. The attractive van der Waals force changes because the medium between the roller and the glass plate changes from air to ink. The approximation, which can be used in this scenario, is the interaction between two plane-parallel plates separated by a medium m of thickness l (see figure 3.5) since the dimensions of roller and glass plate in all directions are much bigger than

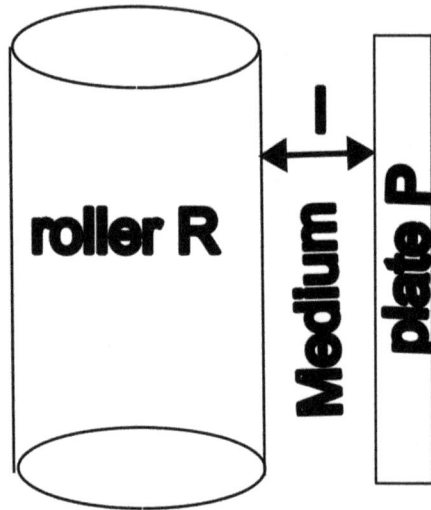

Figure 3.5. The surface of the roller separated by a medium of thickness *l* from the surface of the glass plate.

the separation *l*. When ink is on the roller and glass plate, *l* will have a lower limit of the order of the diameters of the pigment particles in the ink and the roughness of the surfaces involved, that is between 50 nm to several microns.

The electrodynamic work, or free energy $G_{Rm}(l)$, is that required to bring the two plates from infinity to distance *l* as a function of the difference of the dielectric susceptibilities of each plate and the medium in between them. $G_{Rm}(l)$ can be written in the Hamaker form [6]:

$$G_{RmP}(l) = \frac{A_{Rm/Pm}(l)}{12\pi l^2},$$

where $A_{Rm/Pm}(l)$ is the so called Hamaker coefficient which is a function of the distance between the plate and the roller and the dielectric susceptibilities of the three materials involved. For non-conducting materials, which is the case for the inking up of the roller, the non-retarded Hamaker coefficient for two macroscopic phases 1 and 2 interacting across a medium 3 and for a large separation, which is the case in our scenario, the formula can be approximated by [6]

$$G_{RmP}(l, \; T \to 0) = \frac{\hbar c}{8\pi^2 n_m l^3} \frac{n_R - n_m}{n_R + n_m} \frac{n_P - n_m}{n_P + n_m},$$

where *n* are the refractive indices of the roller, medium and plate, $\hbar$ is Planck's constant, and *c* is the velocity of light. Of course, we are working at room temperature and *T* will not go to absolute 0, but since we would just like to demonstrate the order of magnitude for the energies and forces involved, it will do for now. The refractive index of glass is about 1.52, that for rubber 1.519, and the refractive for air is about 1. Black linseed oil based inks will have a refractive index between 1.48 [7], pure linseed oil, and 1.84 [8], pure carbon black, let's assume 1.66. When there is air between the rubber roller and the glass plate, then the free energy is

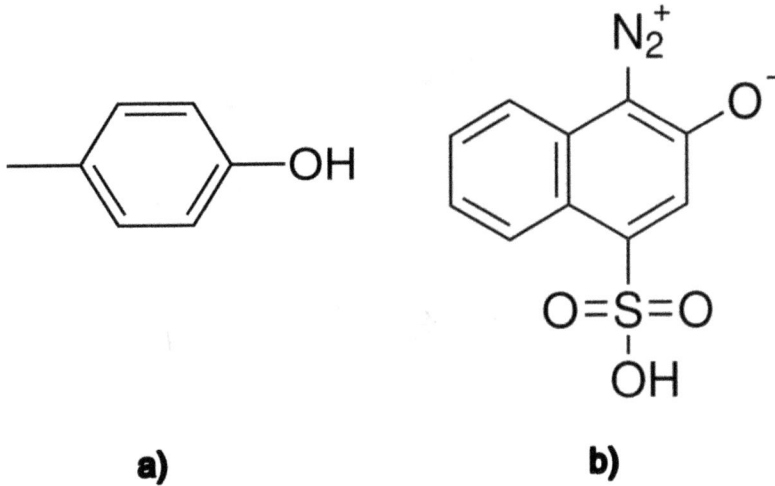

Figure 3.6. (a) The structural formula of p-creosol which is in p-cresol resins and (b) 2-diazo-1-naphtol-4-sulfonic acid ester.

$$G_{RairP}(l,\ T \to 0) = \frac{\hbar c}{8\pi^2 n_m l^3} \times 0.425$$

and for ink in between it is

$$G_{RaikP}(l,\ T \to 0) = \frac{\hbar c}{8\pi^2 n_m l^3} \times 0.002,$$

which is a factor of about 20 and clearly detectable when the ink is rolled out.

Lithography works because water and oil do not mix. A modern photolithographic plate consists of an aluminium substrate and a photosensitive layer. In the case of the Ipagsa ECO 88s plate, the photosensitive layer is 2-diazo-1-naphthol-4-sulfonic acid ester of p-cresol resin which makes the plate light-sensitive to UV light between 350 and 450 nm [9].

Figure 3.6 shows the structure of the two chemical components. Except for the OH group the p-creosol resin has no polar groups. R in the 2-diazo-1-naphtol-4-sulfonic acid ester can be any alkyl chain with the general formula $C_n H_{2n+1}$ or a cycloalkyl with the general formula $C_n H_{2n-1}$.

The smallest alkyl chain is CH_3, methane minus a hydrogen atom, and the smallest cycloalkyl is $C_3 H_5$, cyclopropane minus a hydrogen atom. This surface will be very non-polar.

The plate is a so-called positive plate, i.e. it is exposed to UV light through a film with the positive image. The dark areas, the shadows of the image, where the ink is held, stay as hard elevated areas on the plate, the light areas are washed off during developing in weak nitric acid. The plate is gummed, i.e. a layer of gum Arabic is applied to the plate. Gum Arabic is a complex mixture of different carbohydrate moieties. It consists of 39%–42% galactose units, 24%–27% arabinose units, 12%–16% rhamnose units, 15%–16% of glucuronic acid units, 1.5%–2.6% protein

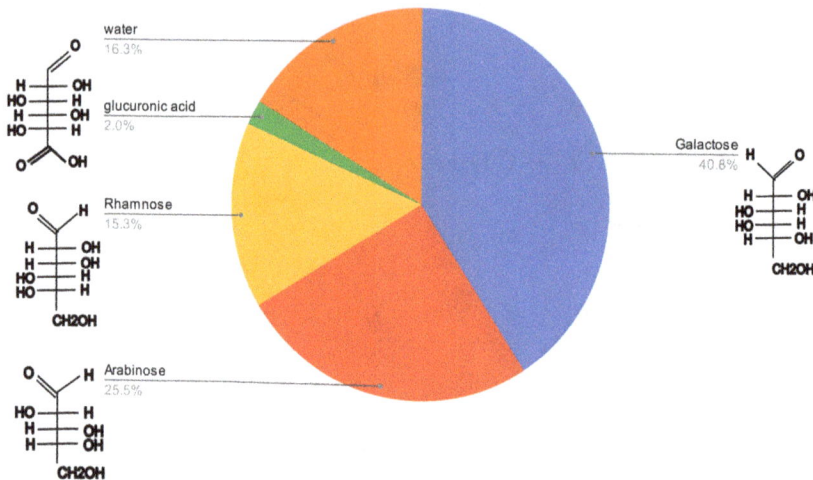

Figure 3.7. The main components of gum Arabic. (From [24].)

moieties, and finally 12.5%–16% moisture (figure 3.7) [11]. The percentage of protein moieties determines the adsorption of gum Arabic species at oil–water interfaces, an important feature for printing applications. As with all natural chemicals, the composition varies depending on where the gum was harvested, how old the Acacia trees were, and what the weather was like.

The hydroxyl groups are polar. The gummed surface is therefore polar. The gum will attach only loosely to the 2-diazo-1-naphtol-4-sulfonic acid ester p-creosol resin surface.

Hydrophobic/hydrophilic interactions are not only important for lithography but for many other printing processes as well. In etching for example, oil-based ink is printed on damp paper. Hydrophobic/hydrophilic interactions in the widest sense determine whether ink can adhere to the substrate.

The hydrophobic/hydrophilic effect is not yet completely understood. For more than 80 years [12] efforts to combine experimental and theoretical results have not lead to a conclusive understanding of the hydrophobic interactions. They remain elusive [13], maybe because water is a special fluid.

The interactions between molecules in a simple fluid can be described by a radial distribution function $g(r)$. It is also called the pair correlation function or pair distribution function [14]. For a simple, isotropic fluid the average density of particles at r with respect to a tagged particle at the origin is [15]

$$\rho(r) = \rho_b g(r),$$

where $r = |r|$, and $\rho_b = \frac{N}{V}$ is the bulk density, the number of all particles in a volume V.

In figure 3.8 the atoms are close together, but not ordered, since the density of a typical liquid is $\rho\sigma^3 \sim 1$ and therefore the probability that the first neighbour shell is found at $r = \sigma$ is high [14]. (r) for a simple, isotropic liquid which obeys the Lennard-Jones

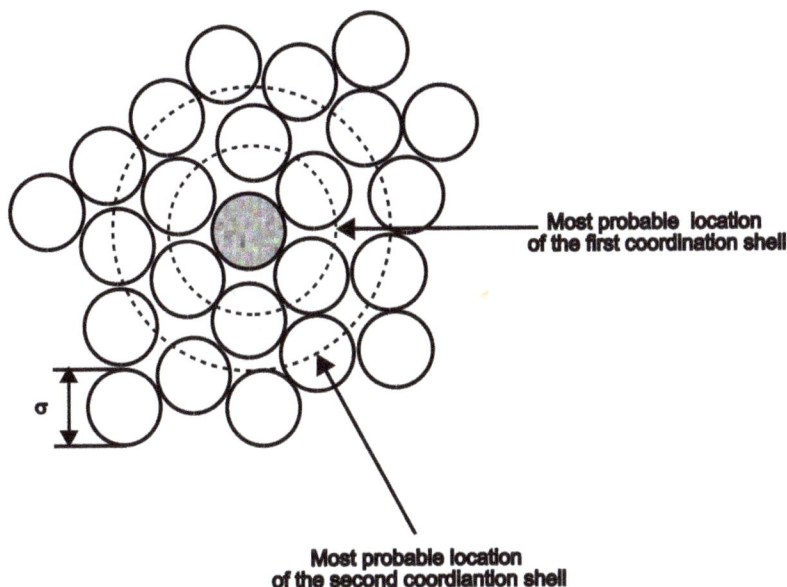

Figure 3.8. Simple liquid structure from [14]. The shaded disc in the middle is the particle tagged at the origin of the coordinate system. σ is the so-called van der Waals diameter, the distance where two atoms or molecules just touch, see figure 3.2.

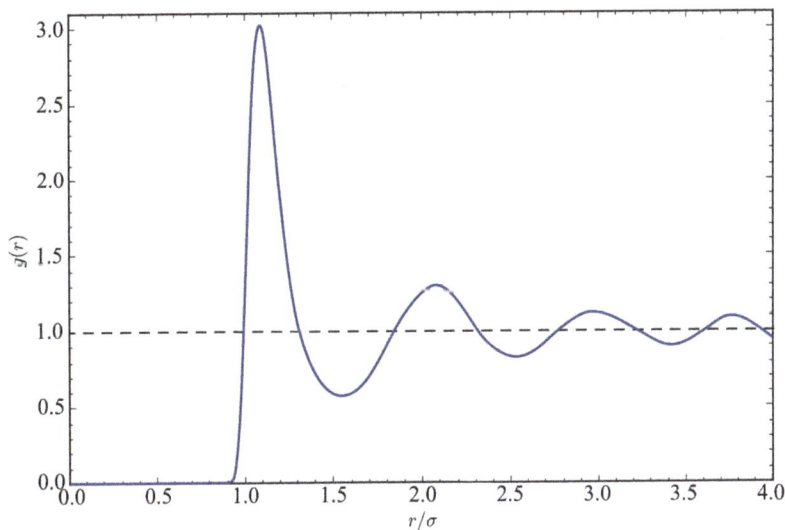

Figure 3.9. Radial distribution function for particles which interact via the Lennard-Jones potential.

potential is shown in figure 3.9. The first peak is at the distance where the first coordination shell occurs, the second peak where the second coordination shell is most likely. Note that the probability to find particles between the first and second coordination shell is not zero. This is typical for a fluid. Whereas figures 3.8 and 3.9

describe liquids consisting of single atoms, water is different, it is H_2O. It consists of two different atoms: H and O and three radial distribution functions: $g_{OO}(r)$ the probability distribution for two O atoms, $g_{HH}(r)$ the probability distribution for two H atoms, and $g_{HO}(r)$ for an H and an O atom. Water is a very polar liquid. The dipole momentum of a single water molecule is 1.84 D. The ionized water molecule has charge of 1.6×10^{-19} °C localised in a diameter $\sim 3 \times 10^{-12}$ m which will cause long ranging perturbation of the arrangement of the water molecules [16]. To complicate the matter, in water hydrogen bonding generates attractive interactions [15]. The hydrogen bond between water molecules leads to a uniquely structured molecular arrangement compared to other known liquids. The three radial distribution functions describe a liquid which shows a tetrahedral order as for ice but, at the same time, a disorder as seen in simple fluids [17].

$g_{OO}(r)$, $g_{HH}(r)$, and $g_{OH}(r)$ can be 'measured' by neutron and x-ray scattering, but as with all measurements of non-crystalline structures, the data are fuzzy and need a lot of interpretation.

The three radial distribution functions are compared to an empirical potential structural refinement (EPSR) simulation. The simulation calculates a three-dimensional structure based on diffraction data. It is based around an atomistic Monte Carlo simulation [18]. The tetrahedral ordering of the molecules in water is the basis of the attempts to explain the hydrophobic effect. Hydrophobic molecules or molecules in general interact in water via different mechanisms [19]:

1. The excluded volume.
2. Attractive forces on water molecules via van der Waals interactions.
3. Strong attractive forces via ionic parts of the molecules, for example charged OH groups.

When the hydrophobic solutes are smaller than a nanometre, the water molecules can maintain their tetrahedral ordering and form large enough cavities to accommodate the non-polar solutes [13], i.e. a few methane molecules (CH_4) or small alcohol molecules (methanol (CH_3) or ethanol (C_2H_5OH)) do not seem to cluster. They do not form any micelles and do not phase separate [20, 21]. They simply exclude the water from the volume of the cavity which is smaller than 0.5 nm [19]. This excluded volume can easily accommodate in the tetrahedral structure of the surrounding water. None of the materials of offset printing belong in this size range.

Table 3.1 shows the composition of linseed oil and the structure of the components. The molecules are not small and are not dissolved into water but are tethered to the surface of the ink layer on the roller. The primary particle size of the pigments dispersed in lithographic ink is a minimum of 11 nm (carbon black) for black or 50 nm for coloured inks. Their surface will be coated with a layer of oil since they are made to be dispersed in non-polar solvents, again the oil molecules are tethered to a surface and are the layer which forms the interface with water. The interaction with water falls therefore into a different regime. When large hydrophobic molecules are inserted into the water matrix, they must break the water structure formed by hydrogen bonds. The hydroxyl groups of the water molecules at the interface point into the hydrophobic layer which results in an orientational ordering of the water molecules similar to that

Table 3.1. Composition of linseed oil from [22].

Component	Mole %	Structural formula
Palmitic acid	7.4	
Stearic acid	5.7	
Oleic acid	23.1	
Linoleic acid	14.9	
Linolenic acid	48.6	

of a liquid–vapor interface [23]. This layer has a lower density than the bulk and is therefore a non-stick layer.

Before the ink is rolled on, the plate is scrubbed with a wet sponge which removes the gum from the dark areas, but not from the light ones. The dark areas have mostly non-polar molecules with long alkane chains at the surface (see figure 3.6). The hydrophobicity of the dark areas becomes visible since the water will not wet these areas but coalesces into small puddles. The gummed areas with their many hydroxyl groups at the surface (figure 3.7) allow the water molecules to attach to the surface without changing the tetrahedral bulk order and therefore hold the water film nicely. Every time the plate is inked, it will be wiped with a damp sponge before the ink is rolled on. By the pressure of the roller on the plate the ink is brought into close contact with the surface of the printing plate. This contact will lead to a liquid–vapor interface in the gummed areas and the oily ink will not attach but stay on the roller. The dark areas have a hydrophobic surface, but a thin layer, maybe a monolayer, of water is still attached to them. Even though the dimensions in [25–27], for example, are those for colloidal particles we think that the same depletion process will take place and the water layer between the ink roller and plate will be expelled.

The adhesion of the ink to the plate surface must be the same or bigger than the cohesion in the ink, therefore at least $4900\,\mathrm{N\,m^{-2}}$. The pressure of the roller makes sure that the plate and the ink on the roller come into close contact and adsorption to the plate can take place.

The ink from the plate is then transferred from the plate to the rubber blanket (figure 3.3) and finally to the final print substrate, most often paper (figure 3.3). During ink transfer, the ink is not detached from plate or blanket, but the ink layer is split. Most paper surfaces are absorbent.

In addition to the adhesion of the ink to the surface capillary forces will draw the ink into the surface. When a dry and porous substrate is brought into contact with a liquid, it will absorb the liquid at a rate which will decrease as a function of time. Inks for impact printing are pastes and therefore the penetration depth is rather small.

When printing a full colour print in process colours (yellow, magenta, cyan, black) the first, yellow, print is overprinted three times, often when the ink of the previous layer has not yet dried. The previous layer will leave a 'ghost' on the blanket and good registration between the different layers is required to achieve good image quality.

All printing is based on splitting the ink layer several times. The split happens in different conditions depending on the printing process to which the ink rheology is adapted.

During intaglio printing, etching, photogravure, etc, for example, the ink is first generously spread on the plate and then wiped off until the ink is only held in the depressions of the plate. The image is then transferred by putting the plate on the print bed, covering it with a damp piece of paper, covering the paper with an elastic felt layer and then exposing it to pressure by rolling a steel cylinder over it. This time we have hydrophobic interaction between the ink and substrate which prevents bleeding of the ink, elastic rolling of a cylinder over a plane and adhesion of the ink to the substrate which has to overcome the cohesion of the ink in the indentation to pull the ink of the plate.

In relief printing, letterpress, linocut, etc, the ink sits on top of the elevations of the plate. The ink is applied by rolling it on, again we have the combination the forces of adhesion, cohesion and rolling of a cylinder over a plane during inking and then the competition of adhesion and cohesion when the print is pulled after it has been exposed to pressure in the press.

In screen printing, the split happens when the ink is pushed through the stencil and happens between the ink attached to the mesh and the layer attached to the substrate. Maybe the only printing method where no split happens but instead a complete transfer of toner to the substrate is non-impact printing based on electrostatic powder transfer.

The next step in the printing process is the drying or curing of the ink on the final substrate.

3.2 The drying of ink on different substrates

Printing ink is defined as a coloured substance which is used to transfer image information from a image carrier to the final substrate [28]. It can be classified into two groups: pastes and liquids. During processing they all start out in the wet phase and end up in the dry phase. Drying is achieved by absorption into the substrate, oxidation, evaporation, energy curing, and heat induced polymerization. The odd ones out are solid ink and dry toners. They do not undergo drying. The wet phase of solid ink is the melt and the dry phase the resolidificated ink after cooling. Toner is never really wet. It is a powder and is attached to the substrate by heat fusion [28].

How ink dries is a function of its ingredients. The ingredients can fall into three categories [28]:

1. *The so-called vehicle or varnish.* This is the substance which carries the colourant from the press/printer to the substrate during printing. The vehicle is typically a liquid and consists of solvent, resins, binders and plasticizer. It is the part of the ink which changes or separates from the rest of the ink during drying and is responsible for shrinkage and absorption.
2. *The colourant.* The colourant makes the print visible. It is either a pigment or a dye. Pigments and dyes are made from the same chemical molecules. The difference between a pigment and a dye is that a pigment does not dissolve in the vehicle, but a dye does. Pigment based dyes are more common than dye-based dyes because of increased light fastness.
3. *Additives.* Additives control certain aspects of the ink, for example the stability of the pigment suspension in the vehicle, the pH, the viscosity, the rub/abrasion resistance of the dried ink, and the drying itself, to name a few.

As table 3.2 shows, most printing processes print on absorbing and non-absorbing substrates. Drying and adhesion after drying is different whether the ink penetrates the substrate or sits on the surface. All absorbing substrates allow the ink to penetrate the surface before other drying processes set in. In table 3.3 the composition of different substrates is listed.

The main difference between absorbing and non-absorbing substrates is that absorbing substrates present a bigger surface which increases drying speed, and they stabilize the ink trace on the substrate by locking the ink into place. Both substrate classes can show problems with so called bleeding, where the ink spreads on or in the substrate beyond the image defining areas. This can even happen in the dry phase given enough time. For non-absorbing substrates bleeding is controlled by adhesion and the compatibility of the ink and surface chemistry. For absorbing substrates, the viscosity of the inks and the porosity of the substrate have to be matched in such a way that the capillary forces in the pores do not lead to uncontrolled spreading. Normally during the penetration process the pigment concentration in the ink rises quickly until the ink gels. This minimizes bleeding, but if the substrate is too porous and the viscosity of the ink too low, it cannot be avoided.

In table 3.4 different ink types and their drying processes are listed. Not surprisingly the most common drying process is evaporation.

Table 3.2. Overview of printing processes, ink types, plates, drying methods, and substrates, based on table 2.1 in [29].

Printing process	Ink	Plate type	Drying method	Substrates
Lithographic offset	Oil based paste, energy curable paste	Planographic	Absorption, evaporation, oxidation, UV curing, electron beam curing	Paper, cardboard, metal, glass, ceramics, plastic; can be used on textiles, but not done industrially
Gravure	Solvent based liquid, water based liquid	Recessed (engraved)	Evaporation	Paper, cardboard, plastic film, aluminium foil, vinyl, laminates, non- porous materials
Intaglio	Oil based paste, solvent based paste, energy curable paste	Recessed (traditionally etched, or engraved)	Absorption, evaporation, oxidation, UV curing, electron beam curing	Similar to gravure
Flexographic	Solvent based liquid, water based liquid, energy curable liquid	Raised image (relief) on pliable material	Evaporation, precipitation, UV curing, electron beam curing	Paper, cardboard, polyethylene film, polyester film, vinyl, cellophane
Letterpress	Oil based paste, solvent based paste, energy curable paste	Raised image (relief) on rigid material	Absorption, evaporation, oxidation, UV curing, electron beam curing	Similar to flexography
Screen	Oil based paste, solvent based paste, energy curable paste, water based paste	Stencil	Oxidation, evaporation, UV curing, electron beam curing	Paper, cardboard, metal, glass, ceramics, plastic, main printing method for textiles
Ink-jet	Water based liquid, solvent based liquid, energy curable liquid	Plateless	Absorption, evaporation, UV curing, electron beam curing	Paper, cardboard, metal, glass, ceramics, plastic, textiles
Electrophotography	Dry powder	Plateless	Heat fusion	Paper

3.2.1 Inks that dry via evaporation

Inks that dry via evaporation consist of a solvent which is to a certain extent volatile, pigments or dyes, co-solvents and dispersants. Drying by evaporation means that at the liquid–air interface the energy of the liquid molecules is so high that the

Table 3.3. Different kinds of absorbing and non-absorbing substrates.

Absorbing substrates	Composition	Non-absorbing substrates	Composition
Paper	Cellulose, with or without a sizing of kaolinite, calcium carbonate, bentonite	Coated paper	Cellulose with a non-absorbing coating for example kaolin clay in a latex binder
Ceramics	Kaolin with different additives like silica or feldspar	Glazed ceramics	Vitrious coat of different origins
Glass	3D printed silica glass	Glass	Silica
		Metal	Iron, silver, gold, copper, palladium, platinum, aluminium, alloys
		Plastic	Organic polymers such as polyamide, polycarbonate, polyester, polyethylene, polypropylene, polystyrene, polyurethane, polyvinyl chloride, polyvinylidene chloride, acrylonitrile butadiene styrene, phenolics or phenol formaldehyde, melamine formaldehyde, polylactic acid, etc
Fabric	Polyamide, polyester, polyacrylonitrile, protein fibre (wool and silk), cellulose fibre (cotton, flax, jute, bamboo, rayon), mineral (asbestos, glass fibre)	Coated fabric	Coated fabric, for example with polyurethane
Leather	Animal protein	Coated leather	For example with polyurethane

attractive forces between the molecules, the cohesion, is smaller than the thermal energy of the solvent molecules and they leave the liquid phase and enter the gas phase. In the case of pigmented inks, the pigments are left behind since they are much bigger particles than the solvent molecules and would require such large thermal energies to enter the gas phase that the print would disintegrate. For inks where the colorant is a dye, the dye molecules need to adsorb to the substrate and their boiling point must be higher than that of the solvent. Inks which dry by evaporation loose almost all their vehicle during the process. Depending on the boiling point of the solvent in the vehicle it occurs at room temperature or heating must be applied to speed up the drying process. Drying by evaporation means that the amount of colourant left on the substrate is only a fraction of the weight of the wet phase of the ink. Evaporation leads to shrinkages, which can cause the deformation of the substrate or leave so called coffee rings on a non-absorbing surface. For example heat set lithographic inks have a pigment concentration of

Table 3.4. List of drying processes for different inks, from [30].

Ink type	Ink chemistry	Drying process[a]	VOC composition	Typical% VOC	On-press emission controls
Paste inks					
Offset sheet-fed	Oleo resinous	Oxidation	Aliphatic hydrocarbons	0–20	None
Offset heatset	Oleo resinous	Evaporation	Aliphatic hydrocarbons	35–45	Afterburners
Offset coldset	Oleo resinous	Substrate absorption	Aliphatic hydrocarbons	2–20	None
Energy curable	Acrylated monomer/ oligomer	Polymerization	Unknown	0–5	Venting to atmosphere
Liquid inks					
Flexo-gravure solvent	Various resin-solvent combos	Evaporation	Various solvents	40–70	Afterburners
Flexo-gravure water	Various resin types	Evaporation	Alcohol (if present)	0–2	Venting to atmosphere
Gravure publication	Resin-toluene	Evaporation	Toluene	40–70	Recapture
Inkjet solvent	Various resin-solvent combos	Evaporation	Various solvents	40–90	
Ink jet water	Various resin types	Evaporation	Water	0–5	None
Solid inks					
Dry toner	Various thermoplastic resins	Fusion via high heat	None	NA	NA
Hot melt	Resin and wax	Melt and resolidify	None	NA	NA

[a] All printing inks also dry by absorption into the substrate, dependent on the porosity of the substrate.

about 10 wt% [28] and another 33–47 wt% of non-volatile additives. That means that about 43%–57% of the weight of the wet phase turns into vapor of hydrocarbons. The hydrocarbons are volatile organic components (VOCs) which are burned during processing, see table 3.4. Note that water-based inks loose only up to 5 wt% of their weight during drying, similar to UV curable inks, and the volatile component is either water or alcohol. They are therefore quite favoured when it comes to controlling VOCs, which is particularly important for the printer market aimed at consumers.

3.2.2 Polymerizing inks

Polymerizing inks are inks which 'dry' by transforming the liquid state into a solid by forming a network. Two types of network formation are used in printing inks: the linking of monomers to form polymers by UV or e-beam curing and the linking of molecules by oxidation. Polymerization by heat is not very common in the printing industry anymore and has been mostly replaced by UV curing because of the lower VOC emissions and reduced energy consumption.

3.2.2.1 UV curing inks

The fast transformation of a photo-polymeric liquid to a polymeric solid is exploited in a wide array of applications such as coatings, adhesives, printing inks and 3D printing. UV curing inks are favoured when low shrinkage, high drying speed and light fastness is desired. Ink formulations have been developed for gravure, flexographic, screen, ink jet and pad printing and are suitable for printing on paper, cardboard, metal, glass and plastic packaging, signs, wallpaper, magazines and books.

A photopolymer is any substance which changes its chemical or physical properties by interaction with light. UV curable inks belong to the type where multifunctional monomers or oligomers undergo a radical, cationic or anionic chain polymerization upon irradiation with UV light and form a cross-linked network [31]. For this kind of photopolymerization a photoinitiator is required which is the source of radical, cation or anion formation when interacting with light. Figure 3.10 shows the three steps of photopolymerization with a photoinitiator.

UV curing systems consist of a combination of the following ingredients [32]:
- Pigments (dyes do not survive irradiation with UV).
- Multifunctional oligomers (prepolymers).
- Multifunctional monomers (diluents).
- Photoinitiators.
- Co-initiators (reducing agents, chain transfer agents, spectral sensitizer).
- Light stabilizers.
- Thermal stabilizers.
- Plasticizer.
- Additives.

The pigments are dispersed in the carrier and 'freeze' in place during polymerization. Photoinitiators, prepolymers and diluents are the main players in the polymerization process. The two main types of initiators are: radical or cationic initiators. The radical initiator is split by homolytic photocleavage into radicals which trigger a chain process in which the prepolymers and diluents are converted into crosslinked, solvent and chemically resistant films. The second type produces a proton acid by photolysis, an electron is split from a molecule and leaves a positively charged proton acid behind, which then initiates the polymerization process [31, 33–35].

Photoinitiator
+ ──▶ **Reactive species**
UV-radiation **+** ──▶ **Crosslinked polymer**
Multifunctional monomer

Figure 3.10. The steps of light induced polymerization. (From [34].)

Radical polymerization is used with acrylates, styrene/unsaturated polyester systems and thiol/polyene systems. Cationic polymerization is used for multifunctional epoxides and vinyl ethers (see table 3.5). Acrylate-based resins are the most popular choice because of their fast reaction speed and their large choice of monomers, but they have to be cured in a protective atmosphere since oxygen interrupts the polymerization. This is not the case for cationic curing, i.e. expoxides and vinyl ethers, as they can be cured under air, but their conversion rate is lower. Independent of the process the rate of polymerization R_p is a function of the concentration of photoinitiator and the light intensity at the distance d from the surface [33, 36]:

$$R_p(d) = \frac{k_p}{\sqrt{2}\,k_t}[M]\overline{\sqrt{\Phi_i I_a(d)}},$$

where k_p is the rate constant for the propagation step, i.e. how fast the monomers attach to each other, typically $10^4 \, l\,mol^{-1}\,s^{-1}$ for acrylate systems [33], k_t is the rate constant of termination, i.e. how fast the polarization runs its course, typically 10^5 $l\,mol^{-1}\,s^{-1}$ for acrylate systems [33], $[M]$ is the monomer concentration, ϕ_i is the initiation quantum yield, and I_a is the absorbed light intensity. The quantum yield or the quantum efficiency is the number of molecules participating in a certain process divided by the number of photons required to trigger the process. For example, for $\phi = 1$ every photon creates a photochemical reaction, for $\phi < 1$ other chemical reactions compete or hinder the photochemical reaction, for $\phi > 1$ a chain reaction is taking place, i.e. more than one chemical reaction is triggered by one photon.

The absorbed light intensity is a function of the concentration of the photo-initiator [PI], the distance from the surface d and the molar absorption coefficient of the photoinitiator ε, and follows the Beer–Lambert law

$$I_{acul}(d) = I_0(1 - e^{-\varepsilon[PI]d}).$$

This is the cumulative absorbed intensity when the light has travelled from the surface to the depth d in the film. The absorbed intensity at d is

$$I_a(d) = I_0\varepsilon[PI]e^{-\varepsilon[PI]d},$$

and with that the rate of polymerization is

$$R_p(d) = \frac{k_p}{\sqrt{2}\,k_t}[M]\overline{\sqrt{\Phi_i I_0\varepsilon[PI]}}e^{-\varepsilon[PI]d},$$

$\frac{k_p}{\sqrt{k_t}}$ is called the characteristic ratio of the monomer system. It is independent of the kind of photoinitiator but temperature and viscosity dependent. It measures the ability of the monomer to support a chain reaction. The termination rate k_t decreases with increasing viscosity, i.e. the polymerization rate is higher in viscous systems [36]. Doubling the intensity of the UV illumination will only increase the polymerization rate by roughly 1.4. The same is true for doubling the concentration of the photo-initiatior. Each UV ink is formulated to hit the sweet spot of the minimal amount of photoinitiator creating the maximum amount of polymerization.

Table 3.5. The two different groups of photopolymerization from [33].

Mechanism	Radical	Cationic
Photoinitior	Aromatic ketone	Aryliodonium salt
Monomers and functionalized polymers	Acrylate	Epoxides

(*Continued*)

Table 3.5. (*Continued*)

Mechanism	Radical	Cationic
	Maleate	Vinyl ethers
	Styrene	
	Thiol/polyene	

3.2.2.2 Electron beam curing inks

The chemistry of UV curing inks and electron beam (EB) curing inks is almost identical except that the EB curing inks do not contain any photoinitiators, which makes them especially suitable for food packaging since photoinitiatiors cannot migrate from the cured ink layer into the food products. During EB curing the substrates stay cold. Therefore, the inks can be applied to substrates which are heat sensitive. The electrons which are generated by an accelerator collide with the molecules of the prepolymers and the diluents in the ink formulation. The collision knocks an electron form the outer shell, ionizes the monomer molecule and initiates an avalanche of secondary electrons. The monomer becomes reactive and starts the polymerization process.

The secondary electrons interact with the molecules in the system until they have lost so much energy that they cannot detach any further electrons. The ions can undergo dissociative electron capture and catalyse further polymerization [36]. In highly pigmented coatings EB curing provides better colour stability and the penetration depth of the electrons is higher than that of UV radiation, allowing thicker, fully cured ink layers. EB curing can only be done under a protective atmosphere since the presence of oxygen hinders curing but also degrades the ink.

3.2.2.3 Drying oil-based inks

Linseed oil has been used in inks for intaglio and relief print since the invention of printing. Modern inks for letterpress and offset lithography are based on drying oils modified with synthetic resins to improve their performance. The drying process is a two-step process. The ink is first absorbed into the substrate and then the polymerization via oxidation sets in.

Polymerization via oxidation relies on the presence of unsaturated molecules, i.e. double and triple bonds. Linseed oil will polymerize in air at room temperature without any catalyst and will continue to do so for many years even after the ink seems to be completely dry [37]. The double bonds of oleic, linoleic and linolenic acid (see table 3.1) react with oxygen and with each other to create a network and form a solid film. The formation of the network is the result of the intermolecular coupling of radicals from the decomposition of unstable peroxide groups [37]. Oleic acid needs temperatures above room temperature to oxidise whereas linolenic and linoleic acid oxidise rapidly at room temperature. The oxidation products are allyl ($-CH=CH-CH_2-$) and hydroperoxide groups ($-OOH$) [37]. Linseed oil does not show any shrinkage but can gain up to 40% of its original weight during oxidation [37]. Linseed oil-based inks are mostly used on porous paper since the polymerization can take a long time and the ink remain as a paste for a considerable amount of time. It can therefore be wiped off non-absorbing substrates easily.

Oxidation drying is slowed down by acidic papers, in particular in high humidity (intaglio is printed on damp paper for example) [36]. The pH values of paper range from 4.9 to 7 for uncoated or rosin sized papers [38] and for alkaline papers between 8.5 and 9.5 [39]. Oxidation drying inks should not be printed on papers with a pH lower than 5 [40].

3.2.3 The odd ones: solid ink and dry toner

3.2.3.1 Solid ink

As the name says, solid inks start as solids and are never wet. They become a liquid just before they are transferred to the substrate or undergo a chemical reaction triggered by heat. In direct thermal printing the ink is part of the substrate. Nothing other than thermal energy is transferred from the print head to the print. The heat sensitive coating of thermal papers consists of a dispersion of colourless colour formers and acidic colour developer in a binder. Typical binders are polyvinyl alcohol, polyvinyl acetate, modified cellulose and modified starch [41, 42]. When heated the acidic developer (phenolic or salicylic acids [41]) melts and comes into contact with the dye precursors (of the triphenylmethane and fluorane type [41]). The acidic developer donates a proton that reacts with the lactone ring of the colour former which then shifts the electron path in the molecule in such a way that colour a resonance is formed in the visible range (figure 3.11). The colour forming process is reversible and stabilizers have to be added to the coating. The most important colour for direct thermal printing is black since the process is mostly used in applications such as label printing, bar code printing, and medical imaging, where colour is not necessary. However, the paper used in pocket photo printers is exploiting a variation of the above process where the crystalline form of the colour former is colourless but the amorphous form after heating is coloured [43].

Thermal transfer printing transfers solid ink from a donor ribbon or sheet to the receiver substrate by melting the solid ink.

In thermal mass transfer or thermal melt printing the donor ribbon is coated in wax or wax resin mixture with the colorant dispersed in it. The coating should have a

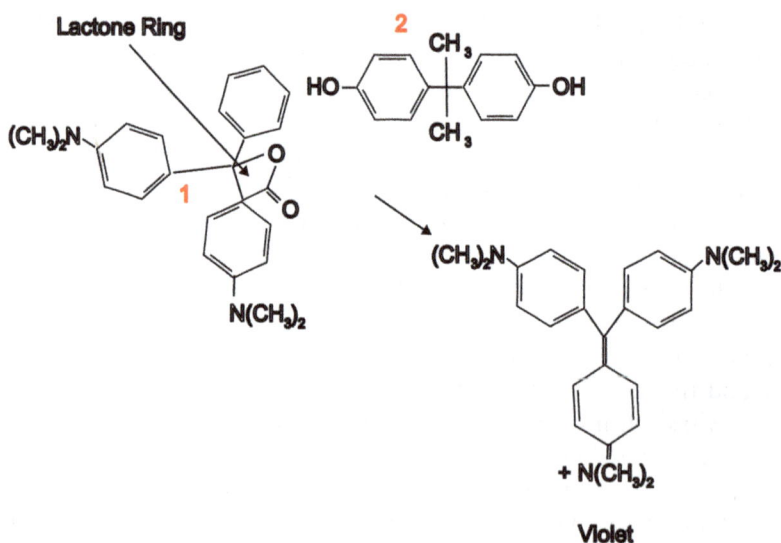

Figure 3.11. Crystal violet lactone (1) accepts a proton from bisphenol A (2). The lactone ring opens and crystal violet forms. (From [42].)

defined melting point with a sharp drop of viscosity at which point the ink is transferred by pressure to the receiver substrate where it solidifies again. The thermal mass transfer receiver substrate can be any material which provides good wetting conditions for the ink [41].

In dye-diffusion thermal transfer (D2T2) or sublimation printing the binder is not transferred to the substrate. Only the dye diffuses into the receiver substrate when it is released from the donor ribbon or sheet by heat and pressure. In a D2T2 donor layer solvent or disperse dyes are dissolved in a binder, for example a cellulose derivate such as ethylhydroxyethylcellulose or a vinyl derivative such as polyvinyl butyral. Polyester is the most common receiver material as it is receptive to solvent or disperse dyes [42].

3.2.3.2 Toner
In electrophotography, as an example for all powder-based printing technologies, a latent image is formed on a charged surface by charging the whole surface first and then discharging it selectively to generate the image information. The image is then 'developed' by using the electrostatic attraction of coloured powders to the charged image areas. The fine powder is then transferred from the image carrier to the final substrate and forms the final print. The toner needs to be fused to the substrate to make it permanent. The toner is not only the colourant but also a resin (40% to 95% of the toner [42]) which facilitates the fusion of the print to the substrate. Heat fusion is the most common method. As with solid ink, the resin needs to have a sharp drop of viscosity upon melting and needs to wet the substrate well to allow the image to penetrate the paper surface for example and become permanent. In most systems, the heat is applied via a heated rubber roller treated with silicone oil to prevent sticking of the toner to the heated roller. In cold fusion the pressure of the cold roller is so high that the toner melts and penetrates the substrate.

3.3 Different formulations for different inks

The two big groups of inks we will consider here are artist's inks and industrial inks. Artist's inks are made in small batches, mostly based on traditional and simple recipes, and made with plant-based solvents or water. Industrial inks are much more complex since they must have colour stability, reproducible performance independent of batch number, extended shelf live, tightly controlled rheological parameters, fast drying, and an as low as possible price point.

When an ink is formulated, a number of factors have to be considered [28]:
- Environmental and health and safety regulations. (In a global market, huge differences still exist between countries and target markets. For example, whether the ink is designed as a consumer product in Europe or as an industrial ink in Asia, will define the set of environmental and health and safety rules which must be followed.)
- Printing process.
- Type of press.

- Press speed (one of the main differences between artistic and industrial applications).
- Type of substrate.
- Drying method.
- Appearance of finished print (for example gloss, transparency, special effects, etc).
- Level of print quality.
- Post press handling.
- End-use requirements, customer base (what is the print used for, who will use it and what is the expected lifespan of the print).
- Cost and ease of handling.
- Cost of raw materials.
- Profit margin of the finished ink.

All inks are made of colorants and vehicles or binders and depending on the applications, they will also contain additives and carrier substances or solvents.

A colorant gives the ink its colour by removing energy from the illuminating light. It is either a pigment or a dye. Pigments are small particles formed by dye molecules which do not dissolve in any of the components of the ink. Pigments require a vehicle or a binder to attach to the substrate and must form a stable colloidal suspension in it. The average primary particle size of pigments is between a few nanometres up to several micrometres. The pigment size determines how much light is scattered by the ink and therefore has an influence on the colour strength, gloss, and opacity or hiding power.

One way to describe the interaction of a particle with light is via cross sections and efficiency coefficients. There are four cross sections for each particle. The most obvious one is the geometric cross section. For a sphere, for example, it is $A_{\text{cross}} = \pi r^2$.

The other cross sections are interaction cross sections. The conservation of energy, i.e. no energy can be destroyed and no energy can be generated from nothing, demands that all the energy scattered from and absorbed by a particle has to be equal to the energy which fell onto the particle. As a convention, this interaction can be described as a function of an interaction cross section which can be smaller or bigger than the geometrical cross section. The extinction cross section is a measure for how much energy is removed from the incident beam of light by a particle and is the sum of the scattering cross section and the absorption cross section:

$$C_{\text{ext}} = C_{\text{sca}} + C_{\text{abs}}.$$

For a completely transparent particle $C_{\text{abs}} = 0$ and the extinction cross section is equal to the scattering cross section. The efficiency factor is a dimensionless factor which describes how efficient a particle of radius r is when it comes to scattering and absorption:

$$Q_{\text{ext}} = \frac{C_{\text{ext}}}{A_{\text{cross}}} = \frac{C_{\text{sca}}}{A_{\text{cross}}} + \frac{C_{\text{abs}}}{A_{\text{cross}}} = Q_{\text{sca}} + Q_{\text{abs}}.$$

The majority of pigments used for inks, even the ones called strongly absorbing, have a very small imaginary part of their refractive index [44]. The imaginary part describes the absorption of a material. The extinction factor will therefore be almost identical to the scattering factor, and we will therefore assume that $Q = Q_{sca}$ in the following. To simplify further we will not discuss the different shapes pigments can have but approximate the shape by a sphere. We will also neglect quantum effects such as fluorescence and assume that all scattering events within an ink layer are independent. All these assumptions allow us to calculate the scattering efficiency factors as special cases of the Mie theory. To do so, we need to decide which approximation can be used by looking at the possible range of particle sizes, the refractive indices of the pigments and those of the media they are dispersed in.

In table 3.6 the refractive indices of pigments used in artist's inks are collected since the refractive indices of industrial pigments are only routinely recorded for white pigments [51]. For extenders see table 3.7. When two or three refractive indices are given, the pigment is birefringent and has two or three axes of different speeds of light. The refractive indices are a function of wavelength and can vary hugely,

Table 3.6. Refractive indices of artist pigments compiled from [45–50].

Colour	Pigment	Refractive index, at $\lambda = 589$ nm
White	Calcium silicate, C.I. Pigment White 28	2.8–3.1
	Titanium dioxide rutile, C.I. Pigment White 6	$\omega = 2.606$–2.616 $\epsilon = 2.899$–2.903
	Titanium dioxide anatase, C.I. Pigment White 6	$\omega = 2.5612$ $\epsilon = 2.488$
	Zirconium oxide, C.I. Pigment White	2.4
	Zinc sulphide, C.I. Pigment White 7	2.37
	Antimony oxide	2.19
	Zinc oxide, C.I. Pigment White 4	2.02
	White lead carbonate	2.01
	White lead sulphite	1.93
	Lithopone, C.I. Pigment White 5	1.84
Yellow	Orpiment, C.I. Pigment Yellow 39	$\alpha = 2.4$ $\beta = 2.81$ $\gamma = 3.202$
	Realgar, C.I. Pigment Yellow 39	$\alpha = 2.538$ $\beta = 2.684$ $\gamma = 2.704$
	Massicot	$\alpha = 2.51$ $\beta = 2.61$

(Continued)

Table 3.6. (*Continued*)

Colour	Pigment	Refractive index, at $\lambda = 589$ nm
	Chrome yellow, C.I. Pigment Yellow 34	$\gamma = 2.71$ $\alpha = 2.29$ $\beta = 2.36$
	Goethite, C.I. Pigment Yellow 43	$\gamma = 2.66$ $\alpha = 2.26$–2.275 $\beta = 2.393$–2.409 $\gamma = 2.398$–2.515
	Cadmium yellow, C.I. Pigment Yellow 37	$\omega = 2.506$ $\epsilon = 2.529$
	Cadmium zinc yellow, C.I. Pigment Yellow 35, 35:1	$\omega = 2.356$ $\epsilon = 2.378$
	Chrome yellow, C.I. Pigment 34	$\omega = 2.31$ $\epsilon = 2.49$
	Lead-tin yellow	Type I 2.29 Type II 2.31
	Lead antimonate yellow, C.I. Pigment Yellow 41	2.01–2.28
	Strontium yellow, C.I. Pigment Yellow 32	$\omega = 1.92$ $\epsilon = 2.01$
	Zinc yellow, C.I. Pigment Yellow 36	1.84–1.9
	Jarosite	$\omega = 1.815$–1.820 $\epsilon = 1.713$–1.715
	Cobalt yellow (aureolin)	1.72–1.76
	Brown spinel, C.I. Pigment Yellow 119	1.719
	Indian yellow	1.67
	Gamboge, Natural Yellow 24	1.582–1.5816
Green	Chrome oxide green, C.I Pigment 18	2.551
	Chrome green, C.I. Pigment Green 15	2.4
	Viridian, C.I. Pigment Green 18	$\alpha = 1.62$ $\beta = 1.62$ $\gamma = 2.21$
	Malachite, Green verditie, Basic Green 4	$\alpha = 1.655$ $\beta = 1.875$ $\gamma = 1.909$
	Emerald green, C.I. Pigment Green 21	$\alpha = 1.71$ $\beta = 1.78$ $\gamma = 1.78$
	Green spinel, C.I. Pigment 50, 26	1.719
	Dioptase	1.65–1.71
	Chrysocolla	$\alpha = 1.575$ $\beta = 1.598$ $\gamma = 1.597$
	Scheele's green, C.I. Pigment Green 22	1.55–1.75

	Volkonskoite	$\alpha = 1.551–1.56$
		$\beta = 1.564$
		$\gamma = 1.569$
	Verdigris, C.I. Pigment Green 20	$\omega = 1.53$
		$\epsilon = 1.56$
	Copper resinate	1.52
Blue	Cerulean blue	1.78–1.84
	Manganese violet, C.I. Pigment Violet 16	$\alpha = 1.67$
		$\beta = 1.75$
		$\gamma = 1.72$
	Cobalt blue, C.I. Pigment Blue 28	1.66–1.74
	Cobalt violet, C.I. Pigment Violet 14	1.626–1.701
	Cobalt zinc blue, C.I. Pigment Blue 72	1.56–1.662
	Lazurite, C.I. Pigment Blue 29	1.5
	Synthetic ultramarine, C.I. Pigment Blue 29	1.51; 1.63
	Ultramarine violet, C.I. Pigment Violet 15	1.56
	Smalt	1.44–1.55
	Prussian blue, C.I. Pigment Blue 27	1.56
	Egyptian blue, C.I. Pigment 31	$\omega = 1.63$
		$\epsilon = 1.59$
	Vivianite	$\alpha = 1.579–1.616$
		$\beta = 1.602–1.658$
		$\gamma = 1.628–1.675$
	Indigo, C.I. Pigment 66	> 1.662
	Azurite	$\alpha = 1.73$
		$\beta = 1.758$
		$\gamma = 1.838$
Red	Hematite	$\omega = 3.13–3.223$
		$\epsilon = 2.87–2.94$
	Vermillion/cinnabar	$\omega = 2.819$
		$\epsilon = 3.146$
	Molybdate orange, C.I. Pigment Red 194	2.55
	Cadmium red (greenockite), C.I. Pigment Red 108, 20, 20:1	$\omega = 2.506$
		$\epsilon = 2.529$
	Red lead	2.42
	Burnt umber, C.I. Pigment Brown 7	2.2–2.3
	Quinacridone red, C.I. Pigment Red 122, C.I. Pigment Red 209	2.02–2.04
	Raw umber, C.I. Pigment Brown 7	1.8–2.17
	Siderite	$\omega = 1.785–1.875$
		$\epsilon = 1.57–1.633$
	Red spinel, C.I. Pigment 33, 35, 12	1.719
	Alizarin (natural), Natural Red 9,6,8,10,11,12	1.7
	Vandyke brown, Natural Brown 8	1.62–1.69

(*Continued*)

Table 3.6. (*Continued*)

Colour	Pigment	Refractive index, at $\lambda = 589$ nm
	Madder lake, C.I. Pigment Red 83, Natural Red 9	>1.66
	Carmine, Natural Red 4	1.6
Black	Iron oxide black, C.I. Pigment Black 11	2.91
	Aluminium flake, C.I. Pigment Metal 1	2.7
	Lamp black, C.I. Pigment Black 6	2.0
	Carbon black, C.I. Pigment Black 7	2.0
	Manganese ferrite black spinel, C.I. Pigment Black 26	1.719
	Chrome iron nickel black spinel, C.I. Pigment Black 30	1.719
	Bone black, C.I. Pigment Black 9	1.65–1.70
	Zinc dust, C.I. Pigment Black 16	1.0402
	Copper powder, C.I. Pigment Metal 2	0.636 60

Table 3.7. Refractive indices of pigment extenders measured at 589 nm, compiled from [52] and [51].

Pigment extender	Refractive index, at $\lambda = 589$ nm
Ground calcium carbonate, C.I. Pigment White 18	1.66
Precipitated calcium carbonate, C.I. Pigment White 18	1.66
Hydrous kaolin. C.I. Pigment White 19	1.56
Calcined kaolin, C.I. Pigment White 19	1.62
Talc, C.I. Pigment White 26	1.60
Crystalline silica, C.I. Pigment White 27	1.54–1.55
Diatomaceous silica, C.I. Pigment White 27	1.45–1.46
Precipitated silica, C.I. Pigment White 27	1.45–1.46
Fumed silica, C.I. Pigment White 27	1.45–1.46
Mica, C.I. Pigment White 20	1.60
Powdered wollastonite, C.I. Pigment White 28	1.55–1.63
HAR	1.63
Feldspar	1.53
Nepheline synite	1.53
Barytes, C.I. Pigment White 22	$\alpha = 1.634$–1.637
	$\beta = 1.636$–1.638
	$\gamma = 1.646$–1.648
Sodium alumino-silicate	1.53
Alumina hydrate, C.I. Pigment White 24	1.57
Barium sulfate, C.I. Pigment White 21	1.64
Calcium sulfate	1.59
Magnesium sulfate	1.57
Aluminium silicate	1.55

especially outside the visible spectrum. Since most printing applications use the visible part of the spectrum and the dispersion across the visible part of the spectrum is on average small, the refractive indices at 589 nm, a standard wavelength for refractive index measurements, are listed.

For a non-absorbing sphere (it was stated before that even for highly absorbing inks the imaginary part of the refractive index is small and can be ignored as a first approximation), the extinction efficiency factor is equal to the scattering efficiency factor and is a function of the radius of the sphere and the difference between the refractive indices of the pigment and vehicle. In figure 20 of chapter 10 in [44] the different regions for approximated efficiency factors are shown as a function of x and n, where x is defined as

$$x = \frac{2\pi r n_{\text{vehicle}}(\lambda)}{\lambda},$$

where $n_{\text{vehicle}}(\lambda)$ is the refractive index of the vehicle as a function of wavelength. Since all the refractive indices listed in the tables are measured at 589 nm, we will drop the wavelength dependency and set $\lambda = 589$ nm. The radius r of the pigments can be between 5 and 5000 nm depending on the application of the ink. Using the smallest and the biggest value for the main component of the vehicle gives the limits for x as 0.0076 and 86. $n(\lambda)$ is defined as

$$n(\lambda) = \frac{n_{\text{pigment}}(\lambda)}{n_{\text{vehicle}}(\lambda)}.$$

When the medium which surrounds the pigment is a vacuum then $n = n_{\text{pigment}}$. Again, the dispersion of n across the visible spectrum will be neglected. The values listed in table 3.6, table 3.7 and table 3.8 give limits for n as 0.89 and 2.24. Zinc dust and copper powder have not been included since they are metals and do not fall in the non-absorbing regime. Industrial pigments are mostly based on organic molecules with $0.89 < n < 1.33$.

In the following three examples will be discussed which should give a good insight in the origin of hiding power.

1. *Transparent inks: Rayleigh–Gans scattering*

When $n - 1 \ll 1$ and $(n - 1) \ll 1$, which means that $r \ll \frac{\lambda}{2\pi(n_{\text{pigment}} - n_{\text{vehicle}})}$ then the Rayleigh–Gans [44] approximation applies and

$$Q_{\text{sca}} = |n - 1|^2 \varphi(x)$$

with

$$\varphi(x) = \frac{5}{2} + 2x^2 - \frac{\sin(4x)}{4x} - \frac{7}{16x^2}(1 - \cos(4x)) + \left(\frac{1}{2x^2} - 2\right)(0.577 + \ln(4x) - \text{Ci}(4x)).$$

From the equation is clear that when the refractive indices of vehicle and pigment are matched, there is no scattering. The hiding power of the pigment is small, independent of its size and the pigment is called transparent. For

Table 3.8. Refractive indices of the main component of vehicles, compiled from [50, 52–55].

Main component of vehicle	Refractive index at 589 nm
Drying oil	
Cottonseed	1.465
Dehydrated castor	1.481
Fish	1.485
Linseed	1.478
Oiticica	1.510
Safflower	1.474
Soybean	1.473
Sunflower	1.473
Tung	1.517
Monomers for radiation curing inks	
Acrylates	1.46–1.48
Expoxides	1.58–1.61
Vinyl ethers	1.428–1.463
Maleates	1.441–1.469
Styrenes	1.5188

carbon black in linseed oil, the refractive index difference is 0.522 and therefore the radius of the particles must be much smaller than 180 nm for Rayleigh–Gans to apply. This can be achieved since the primary particle size of carbon black is about 3 to 6 nm. The scattering efficiency factor will be much smaller than 1 and the ink will display a deep black.

When $x \gg 1$ and $(n - 1) \ll 1$, which means that the particles are rather big, but are almost index matched with the vehicle, which is for example the case for Lazurite in linseed oil, (x) simplifies to $2x^2$. Even for Lazurite particles with a radius of 2500 nm the scattering coefficient is still very small and the ink will have a hiding power which is almost only a function of the absorption by the pigment.

2. *Rayleigh scattering*

Looking at a case where $n - 1$ is neither small nor large, for example calcium silicate in vinyl ether, the scattering efficiency coefficient can be described by the Rayleigh formula when $x \ll 1$ and $(n - 1) \ll 1$.

For calcium silicate in vinyl ether $(n - 1) = 1.171$. $r \ll \dfrac{\lambda}{2\pi(n_{pigment} - n_{vechicle})}$ holds true when $r \ll 56$ nm. Then the formula [44]

$$Q_{sca} = \frac{8}{3}x^4 \left(\frac{n^2 - 1}{n^2 + 2} \right)^2$$

holds true and gives a $Q_{sca} \ll 1$. The ink would be transparent again. To generate pigment particles with a radius much smaller than 50 nm is non-trivial and is mostly achieved for in the industrial production of carbon black or silica particles.

3. *Mie theory*

For arbitrary x and $n \leqslant 2$, which is the case for the pigments and vehicles listed in table 3.6 to table 3.8, the scattering efficiency coefficient for a sphere with real refractive index is described by another special case of the Mie theory [44]

$$Q_{ext} = Q_{sca} = 2 - \frac{4}{\rho} \sin \rho + \frac{4}{\rho^2}(1 - \cos \rho)$$

with $\rho = 4\pi \frac{r}{\lambda}(n(\lambda) - 1)$. λ is the wavelength of the incoming light. Again, only the efficiency coefficient at $\lambda = 589$ nm will be used.

Figure 3.12 shows the behaviour of the efficiency factor as a function of pigment radius and its refractive index. The bigger the refractive index difference between the pigment and vehicle the larger the hiding power. The blue curve is an example of calcium silicate, C.I. Pigment White 28, in vinyl ether. The refractive difference is 1.67 and the relative refractive index of the pigment in the vehicle is 2.17. White pigments do not absorb. All their hiding power comes from scattering. In this case here, even for particles with a radius as small as 70 nm the efficiency factor is already above 1, i.e. the scattering cross section is bigger than the physical cross section of the particle. At a radius of 150 nm, approximately a quarter of 589 nm, the particle scatters most efficiently, the hiding power is at its maximum. The whitest white is

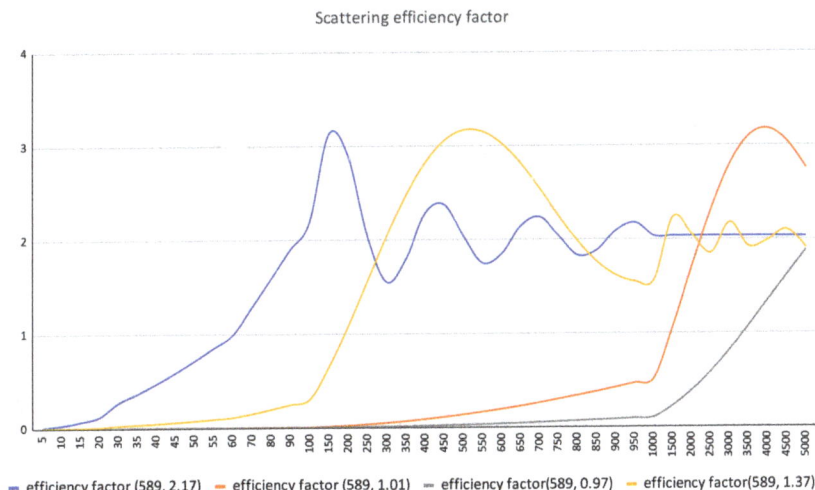

Figure 3.12. Scattering efficiency factor as a function of pigment radius. The blue line is an example for calcium silicate, C.I. Pigment White 28 in vinyl ether, the yellow line for industrial pigments, including carbon black or lamp black, in acrylates, the orange line for lazurite in linseed oil, and the grey line for fumed silica, C.I. Pigment White 27, in dehydrated castor oil.

achieved. For particles with a radius bigger than 1000 nm the efficiency factor plateaus at 2, 30% less than at 300 nm.

Between 150 and 1000 nm the efficiency factor oscillates and falls below 2 for radii about 250, 550 and 850 nm, but stays above 1. The yellow curve in figure 3.12 represents industrial pigments in a UV curing monomer. The refractive indices of industrial pigments are not known in general and are difficult to measure since they are highly absorbent, even though the imaginary part of the refractive index is still very small. Since they are mostly carbon-based chemicals, it is feasible to assume that the real part of their refractive index will not be too different from that of carbon black or lamp carbon, which is 2 (see table 3.6). The most scattering occurs when the pigment radius is about 550 nm, the centre wavelength of the visible spectrum. Scattering becomes small for a radius smaller than 100 nm. Transparent inks, in the industrial sense, therefore always contain pigments with a radius smaller than 100 nm.

Figure 3.13 shows how the oscillation of the efficiency factor changes as a function of the refractive index difference between the particle and medium. For the medium, we assumed a refractive index of 1.5. The blue curve in figure 3.13(a) represents an air bubble which has a refractive index of 1, i.e. the refractive index difference is 0.5. The same refractive index difference leads to the grey curve in figure 3.13(b). This time the particle refractive index is 2, carbon black for example. Both scenarios lead to identical behaviour of the efficiency factor, independent of whether the refractive index of the particle is smaller than that of the medium and the light travels faster through the particle than through the medium, or whether the refractive index of the particle is bigger and the light travels more slowly through the particle. The frequency of the oscillation becomes smaller the closer the refractive indices of the particle and medium are until it disappears, and its amplitude becomes 0 when medium and particle have the same refractive index. The hiding power is then a function of absorption only.

When the colourant is a dye, scattering does not occur even when the dye, the solute, has a refractive index very different to the medium, the solvent. The refractive index of the solution is a function of the components and is described by (from [56])

$$n_{solution} = (N_{solute} n_{solute} + N_{solvent} n_{solvent})/(N_{solute} + N_{solvent}),$$

where n is the refractive index and N is the number of molecules. For dyes and for pigments the absorption will follow Beer's law [57]:

$$I_2(l_2, \alpha) = I_1(l_1, \alpha)e^{(l_2 - l_1)\alpha},$$

where I_2 is the intensity of the light at the depth l_2 as a function of the intensity I_1 at a depth l_1. Often I_1 is the intensity at the surface of the medium and is called I_0. α is the absorption coefficient and is defined as

$$\alpha = 2k_o n_{solution} \kappa_{solution}$$

with $k_o = \frac{2\pi}{\lambda}$, the free-space wave number and $\hat{n} = n(1 + ik)$, the complex refractive index. n is the real part of the refractive index and k the attenuation index, the

a) efficiency factors for n<1

— efficiency factor (589, 0.67) — efficiency factor (589, 0.8) — efficiency factor(589, 0.93)

b) efficiency factors for n>1

— efficiency factor(589, 1.07) — efficiency factor(589, 1.2) — efficiency factor(589, 1.33)

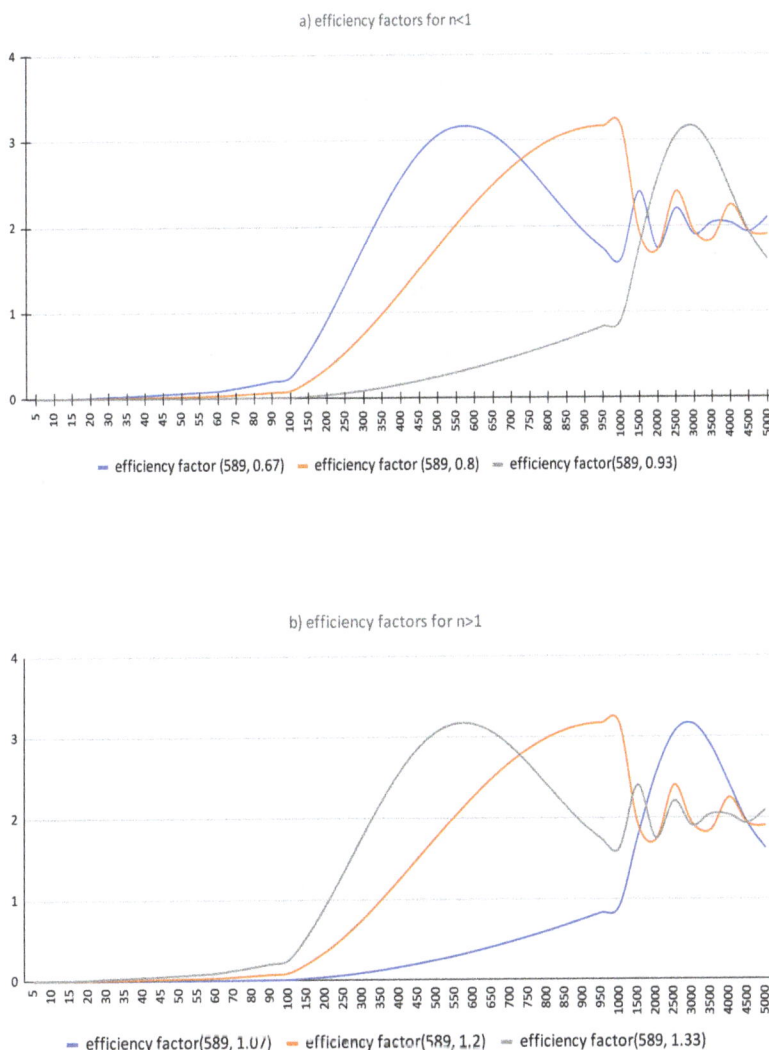

Figure 3.13. For a medium with refractive index 1.5 the efficiency factors for particles with (a) refractive indices smaller than 1.5 and (b) with refractive indices bigger than 1.5 are calculated as a function of the particle radius.

imaginary part of the refractive index, which is small for all non-metallic pigments and was therefore neglected in the scattering formulae. It becomes important when there is no scattering and determines the hiding power of a dye-based ink or in the case of a pigmented ink when it is transparent.

In the following we will list and compare artist's and industrial inks.

Lithographic inks

Lithography is a planographic process and works on the basis that the image on the plate has hydrophobic (the shadows) and hydrophilic (the highlights) areas.

The hydrophobic parts to the plate take the ink and the hydrophilic parts stay ink free. Traditional lithographic inks are therefore oil based. A typical lithographic ink developed for artists, for example, [58] consists of:

- 36 wt% of pigment.
- 62 wt% of linseed stand oil (15–20 Pa s).
- 1 wt% of wax.
- 1 wt% of antioxidant.

Homemade inks are just pigment and linseed oil.

As an industrial printing process lithography is identical to offset printing, a very ink efficient process. The film thickness transferred from the blanket to the substrate is only between 0.5 and 1.5 μm [59], which means that the pigment content has to be rather high to give enough coverage. Offset printing inks are paste-like with a dynamic viscosity of 40–100 Pa s. Conventional offset inks are oil based, but energy curable inks are widely used as well. The ink formulation is determined by the [28]:

- Press design.
- Press speed.
- Plate type.
- Dampening system.
- Fountain solution.
- Blanket.
- Substrate.

As with all commercial inks, the recipes are trade secrets and highly protected. They have many more ingredients than artist's inks and are highly tuned to the application and to guarantee the same performance batch after batch. In industrial inks linseed oil has been replaced by different grades of hydrocarbon oils, driers are added and additives which control the rheology of the ink. An example (from [60]) is the following ink formulation for sheet-fed offset printing on paper:

- 15–20 wt% of pigment and filler.
- 20–30 wt% of hard resin (binder).
- 8–12 wt% of alkyd resin (drying oil, binder).
- 10–20 wt% of triglyceride (drying and semi-drying oil).
- 15–25 wt% of mineral or oil esters (non-drying oils).
- 3–5 wt% of additives, eg. wax, driers, antioixidants.

In UW curing inks the oils and the binder are replaced by a mixture of monomers and photoinitiators.

Intaglio inks

Artist's inks for etching, drypoint and photogravure are very similar to lithographic inks. The main difference is that they do not contain wax and antioxidants. An example is the recipe given by [61]:

- 50 wt% of black pigment (1/3 French Black, 2/3 Frankfurt Black).
- 25 wt% of medium copperplate oil.
- 25 wt% of light copperplate oil.

Depending on the temperature in the room and how deep the marks on the plate are, the viscosity of the ink is adjusted by changing the percentage of the copperplate oils.

The industrial embodiment of intaglio printing is gravure or rotogravure printing. The plate/cylinder is covered in cells of different sizes and depths to vary the amount of ink transferred to the substrate. The cylinder is directly immersed into an ink reservoir and then wiped by a doctor blade removing the ink from the raised areas which are printing the highlights [28]. Gravure inks have a low pigment content and low viscosity and a high solvent content.

The basic formulation from [28] is:
- 4–12 wt% of pigment.
- 0–8 wt% of extender pigment.
- 10–30 wt% of resin.
- 40–60 wt% of solvent.
- 2–10 wt% of plasticizer/wax/additives.

The basic formulation is adapted to the substrate and end use and can have many more ingredients.

Relief inks

Inks for relief print or letterpress are again very similar to lithographic inks, but the ink layer is much heavier than for lithography. The particle concentration is the same as for litho inks, but the pigment concentration is lower. Dryers are added to speed up the drying speed and wax to increase rub resistance. A recipe for a single pigment relief ink is, for example [58]:
- 26 wt% of pigment.
- 10 wt% of extender.
- 62 wt% of linseed stand oil (15–20 Pa s).
- 1 wt% of wax.
- 1 wt% of manganese driers.

The industrial equivalent of artists' relief print is flexography. Flexoplates are pliable and malleable and therefore suited to substrates where the plate needs some give, for example for rough surfaces such as corrugated board. In flexography the ink is transferred from the plate to the substrate. The plate is inked using a so-called anilox roller, a roller covered in ink-holding cells, similar to the rotogravure cylinder but without image information and a regular cell coverage [28]. The inks are low viscosity inks. Depending on the application they can be solvent based, water based or energy curable. Flexography is widely used for packaging. In particular, for flexible packaging the inks must stay flexible to prevent cracking. A large number of containers printed with flexography are laminated which means that the inks have to be compatible with the lamination process. In [28] several general recipes for flexo inks are given. As an example, a recipe for a water based ink for a non-absorbing substrate is given:
- 18–22 wt% of pigment.
- 30–60 wt% of acrylic emulsion.

Table 3.9. Examples of solvent based flexographic inks.

	For polyethylene film	For folding carton
Colorants	12 wt% of organic pigment	14 wt% of organic pigment and 6 wt% of titanium dioxide as filler
Binder	22 wt% of polyimde, 4 wt% of nitrocellulose	11.5 wt% of nitrocellulose, 8 wt% of maleic resin
Solvents	34 wt% of n-propyl alcohol, 13 wt% of ethyl alcohol, 12 wt % of n-propyl acetate	25 wt% of ethyl alcohol, 25 wt% of n-propyl alcohol, 10 wt% of n-propyl acetate
Additives	2 wt% of polyethylene wax, 1 wt% of amide wax	5 wt% of plasticizer, 3.5 wt% of polyethylene wax

- 10–20 wt% of resin varnish solution.
- 4–6 wt% of polyethylene wax compound.
- 2–4 wt% of isopropyl alcohol.
- 15–20 of wt% water.
- 1–2 wt% of amine.
- 0.1–0.5 wt% of defoamer.

The recipe demonstrates that the water-based ink does not contain a huge amount of water. Their environmental advantage is that the amount of VOCs is very small, here 2–4 wt% isopropyl alcohol.

An example of solvent borne flexographic inks from [60], shown in table 3.9, demonstrates the variation of the ink formulation as a function of substrate.

Screen printing inks

Screen printing means pushing a pigment suspension through a screen. The printing process is most versatile, and the ink composition is dictated by the application and the substrate.

Non-impact inks

Non-impact printing is the youngest printing technology, and is without the need of a printing plate. Non-impact printing is the umbrella term for electrophotography, iconography, magnetography, inkjet, thermography and electrography. The inks are specially formulated for different printing methods and for different applications. Non-impact printing is an industrial process and has no 'hand printed' version. In general ink formulations are trade secrets and highly guarded. The two commercially most important methods are inkjet and electrophotography.

    1. *Ink jet*

        Independent of the technology, whether it is continuous or drop on demand, the inks have to fulfil the following requirements [28]:

- The ink must not clog the nozzle. It should not dry in the nozzle including when the nozzle is not used and should not block the nozzle during printing.
- When on the substrate, it has to dry quickly, usually within 1 or 2 s.
- It must lead to an acceptable print quality which is independent of the substrate.
- It must be suitable for the end use application.

Inkjet inks have very low viscosity, mostly smaller than 1 mPa s. In water based, solvent based, and hot melt inks the colorant can be a dye or a pigment. For UV curable systems the colorant has to be a pigment since a dye will not survive the UV curing process. Because of the small size of the nozzle, the pigments have to be so-called transparent pigments with a radius of 80 nm or less. Compared with all other inks discussed so far, the dye or pigment content of the ink is small, between 2 and 5 wt%.

The first general recipe is from [60] and is for a solvent borne ink and the second one from [62], shown in table 3.10 is for a UV curable ink. Both show the complexity of inkjet inks.

Solvent borne ink:

- 42 wt% of methanol.
- 30 wt% of ethyl ketone.

Table 3.10. Features of UV curable inks from [62].

	Alternative name	Cyan %	Magenta %
SR339	PEA	33.15	27.45
SR506	IBOA	16.00	19.00
Ebecryl	8402 urethane acrylate	8.00	
Genocure 1122	Urethane acrylate	18.00	20.00
V-Cap	N-vinyl caprolactam	9.50	9.50
SR9003	PONPGDA	3.10	10.50
BYK-361N	Polyacrylate	0.50	0.50
BYK-377	Polyester modified polydimethyl siloxane	0.050	0.050
Genorad 16	Stabilizer		
Irgastab UV22	Stabilizer	1.00	
Genocure TPO	Acylphosphine oxide, MAPO	9.00	9.00
Cyan	Cyan pigment	1.70	
Magenta	Magenta pigment		35
	Total	100,00	100.00
	Viscosity, cps@45 C	12.5	11.9
	Surface tack	5	5
	Scratch resistance	5	5
	Flexibility	170%	150%

- 1.5 wt% of water.
- 9 wt% of ethyleneglycol methyl ether.
- 1.4 wt% of methyl ester of rosin.
- 13 wt% of styrene-acrylic acid co-polymer.
- 2 wt% of dye.
- 4 wt% of nonyl-phenoxypolyethoxy ethanol.

For the consumer market inkjet inks have to be as environmentally friendly and as versatile as possible. They are solvent free and adhere to paper and non-absorbing surfaces. For the manufacturing market they are tuned to the requirements of the end application which determines the solvents and additives used in the formulation.

2. *Electrophotography*

Electrophotography is another plateless printing process. A photoconductive belt or drum is charged and selectively discharged by exposing it to a controlled flash of light which contains the image information, that is the light and dark areas. The areas of the drum which are exposed to light discharge, the ones which were not exposed keep the charge. A latent image has been generated. The charged parts of the drum, the shadows of the image, attract oppositely charged toner particles which are then transferred to the substrate, mostly paper. The toner particles are fixed to the substrate by heat. The inks for this process are either dry powders or powders dispersed in a liquid. Toners are pigment based. Because they do not need to go through a nozzle, the primary particle size in dry toners is much bigger, between 4 and 10 μm [63]. The toner contains pigment, resins, charge modifiers and additives such as surfactants. As an example for a dry toner the composition patented in [63] is given:

- 80 wt% of latex (resin).
- 12 wt% of wax (releasing agent).
- 3.5 wt% of pigment (colorant).
- 2 wt% of charge control agent.
- 2 wt% of surfactant.
- 0.5 wt% of hydrophobic silica.

The size of the colorant in liquid toners is smaller than for dry toner. It lies between 20 nm and 3 μm [64–66]. Examples of liquid toner compositions can be found in a multitude of patents. As an example, a formulation from [65] is given:

- 2 wt% of dispersant.
- 73 wt% of white oil.
- 20 wt% of epoxy resin.
- 5 wt% of pigment.

The substrate for toner based printing is mostly paper.

3.4 Rheology of ink

All pigmented inks are suspensions or dispersions. Their behaviour under shear load, that is under the force which makes the material slide, is of critical importance for ink applications. The performance of the ink in any printing process depends on its rheological characteristics optimised for the printing process.

Rheology is the study of the deformation and flow of matter under the influence of an applied stress. Flow is defined as continuous irreversible deformation of matter and is a function of the internal and external friction the flowing matter experiences. Viscosity is the retarding influence of a stationary layer on a moving parallel layer [67] and is an internal friction. When shear stress τ, that is the force per unit area, is applied to one layer in a liquid, all adjacent layers will move at velocities which will decrease with increasing distance from the first layer. The velocity gradient is called the shear rate γ. Newton's law describes that the shear rate is directly proportional to the applied force [67]:

$$\tau = \eta \frac{dv}{dx} = \eta\gamma$$

with η the viscosity coefficient. For so called Newtonian liquids, such as water for example, η is constant. For non-Newtonian fluids this is not the case and the viscosity coefficient is a function of the shear stress as well. For complex fluids, for example colloidal suspensions or liquid crystals, the viscosity coefficient can also change with the direction of the shear stress. For all liquids the viscosity is temperature dependent. In general, the viscosity coefficient goes down when the temperature increases.

Figure 3.14 shows the different flow behaviour of liquids.

1. *Newtonian*

   As mentioned before the viscosity coefficient is constant and therefore the shear stress increases linearly with the shear rate. In printing this behaviour can be observed in ink jet, flexo and gravure inks.

2. *Dilatant*

   The term dilatant describes systems where the volume and/or the viscosity of the liquid increases with increasing shear rate. An increase in viscosity can occur without an increase in volume and the phenomenon is then called either rheological dilatancy or shear thickening. A classic example for this behaviour is starch suspended in water. A sudden impact will make the liquid behave like a solid. This behaviour is undesirable in inks but can occur when the pigment load is high and the dispersant is of low viscosity. Time dependent dilatant behaviour is called rheopecty or rheopexy. The viscosity increases with increased duration of a constant shear rate. For example, the viscosity increases with the duration of the shaking of the sample. In lubricants this is a desirable behaviour, but not in inks.

3. *Pseudoplastic*

   Pseudoplastic behaviour is the opposite to dilatant behaviour. With increasing shear rate the shear stress decreases, that is the system flows more easily. Inks and paints are designed to show this behaviour. This allows to the ink or paint to be applied easily and inhibits bleeding after application. The time-dependent variation of shear thinning is called thixotropy. The longer the fluid is exposed to shear the less viscous it becomes. A classic

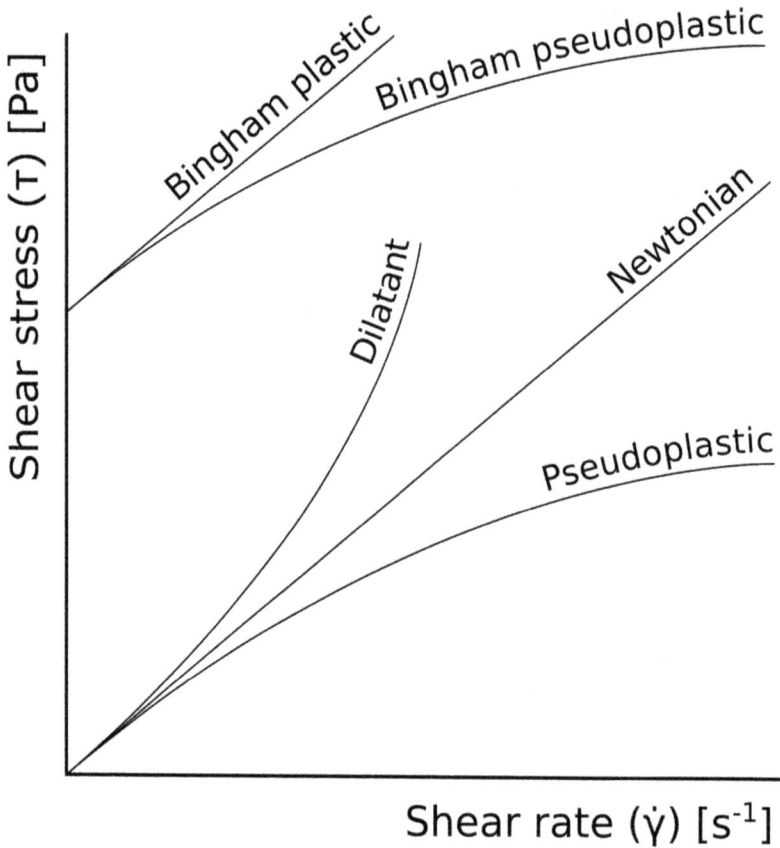

Figure 3.14. The different flow behaviours of liquids. (This image has been obtained by the author from the Wikimedia website where it was made available by g-sec (2013) under a CC BY 3.0 licence. It is included within this chapter on that basis. It is attributed to g-sec.)

example of this behaviour is ketchup. Inks can show time-independent and time-dependent shear thinning at the same time.

4. *Bingham plastic*

At zero or low shear rate a Bingham plastic behaves like a rigid body. After the so-called yield stress is reached, the material starts flowing. If the material flows like a Newtonian liquid then it is called a Bingham plastic [68]. Examples are mayonnaise and toothpaste.

5. *Bingham pseudoplastic*

If the material flows and shows shear thinning, then it is a Bingham pseudoplastic. Bingham plastic and pseudoplastic are important for 3D and 2.5D printing when the print should keep its shape in a third dimension.

Most inks show non-Newtonian behaviour. The measurement of the non-Newtonian viscosity coefficient is non-trivial. Different shear rates and different measurement set-ups will lead to different values. The following devices are used in ink manufacturing and printing:

1. *Orifice viscometers*

   This is a very simply set up, mostly used on the press side for gravure and flexo printing, i.e. for relatively low viscosity inks [28]. The flow time of a fixed volume through a orifice into a measuring cylinder is measured [67]. Another method is the dip method [28]—a cup with a central hole is immersed into a container with ink held at a constant temperature. The cup is pulled out swiftly and the time until the cup has emptied is measured from the moment the cup has completely broken the surface. The viscosity is listed in seconds. The cups come with different hole sizes to accommodate different viscosities. Thixotropy affects the reproducibility of the cup method. It is therefore used most successfully for liquids with Newtonian or near-Newtonian behaviour. It is a hands-on approach which allows press side quality control but not more.

2. *Falling (rising) sphere viscometers*

   The group of falling (rising) sphere viscometers determines the viscosity of a fluid by the terminal velocity of a sphere falling (rising) in the fluid [67]. A wide range of viscosities can be measured using this method. Falling sphere viscometers measure the time a sphere needs to fall between two marked lines in a tube. This time is then compared to the times for standardized control fluids. The bubble tube measurement is another version. This time the bubble/sphere rises. A test tube is filled with a liquid and an airspace is left at the top. The tube is sealed, temperature equilibrated and then inverted. The time it takes for the air bubble to rise to the surface is compared with standardized viscosity fluids [28]. Both methods work only for transparent or translucent liquids where the sphere is visible.

3. *Coaxial viscometers*

   The coaxial viscometer has concentric cylinders. The outer one is called the cup and the inner one the bob or spindle. In most modern concentric cylinder viscometers, the cup is stationary and the bob rotates. The sample is sheared in the gap between the cup and the spindle. The twist of the wire or rod the spindle is hanging on indicates the torque and is used to calculate the shear stress. In a measurement the speed of the spindle is varied and with that the rate of shear. The shear stress is measured and plotted against the shear rate. Analysis of rheological data for non-Newtonian liquids is difficult. Corrections are necessary for shear thickening, shear thinning, yield stress and slip [67]. Nevertheless, data measured with a coaxial viscometer are more reproducible than the data measured with the orifice viscometer or the falling sphere viscometer.

4. *Falling rod viscometer*

   The falling rod viscometer is used for paste inks. A rod is inked and moved up and down through an opening until the sample is evenly applied and thixotropic structures are broken down. The rod is released, and the time needed to fall a defined distance is measured. The weight of the rod can be

Table 3.11. Viscosity of inks for different printing processes from [69].

Printing method	Viscosity range (mPa·s)
Thermal inkjet	1–5
Piezo inkjet	5–30
Gravure	50–200
Flexography	50–500
Screen	1000–10 000
Offset	40 000–100 000

changed. The data recovered from the method are viscosity, yield value and shortness (yield divided by viscosity).

5. *Cone and plate viscometer*

    Another viscometer suitable for paste inks is the cone and plate viscometer. It consists of a flat plate in combination with a low angle ($\leqslant 4°$) rotating cone. The shear rate of the cone and plate viscometer does not change with increasing distance from the axis of rotation since the velocity and the gap increase with distance from the axis. It is the best option for non-Newtonian fluids, i.e. for most inks, since no complicated corrections are needed for non-Newtonian systems [67]. Because of the rather large sample surface, a loss of solvent and therefore a change in viscosity and heating of the sample by friction at high shear rates with the accompanying change in viscosity can be a problem. As a coaxial viscometer the cone and plate viscometer can be run with small sample amounts. Table 3.11 gives examples of typical ink viscosities for different printing processes. Offset inks display by far the highest viscosities. They are classified by another rheological feature—tack.

6. *Tackmeter*

    Tack is defined as the force required to split a single film of ink into two. It is a measurement of the cohesion of the ink in combination with its adhesion to different substrates, surface tension, viscosity and yield value [70]. Tack must be tuned in such a way that 'picking' is avoided and 'trapping' obtained. Picking occurs when the ink is too tacky and the surface of the substrate, especially of paper and cardboard, is ruptured. Small amounts of fibre or coating are picked up by and accumulate on the blanket. Trapping describes the retention of one ink by another. For example, when magenta is printed on yellow, yellow must retain the red ink and pull it off the blanket as the blanket leaves the paper, ideally without staining the blanket yellow.

A simple test for tack is the so-called finger tap out test still used by practitioners. A small amount of ink is spread with the fingertip on a glass plate. The finger is then pressed down and pulled away quickly. This motion is repeated several times and a sensation of stickiness arises. Based on this sensation, an experienced print maker

Table 3.12. Tack range for different printing inks and substrates from [28].

Type of ink	Substrate	Tack range at 1200 rpm
News ink	Newsprint	4–6
Heatset	Super calendered paper	7–10
Heatset	Light weight coated paper	11–14
Sheet-fed	Coated paper	15–18
UV	Coated paper	18–20
UV	Non-porous substrates	19–21
Metal deco	Aluminium or other metal	12–16

can decide whether the tack of an ink is suitable for the printing process. The original tackmeter was based on the same principal [70]. Modern tackmeters measure tack as a function of film thickness, temperature and roller speed. Such tackmeters consist of a set of rollers. The first one is driven and covered by a film of ink of known thickness. The second one is the measuring roller and it is connected to a strain gauge. The third roller helps to keep the ink film evenly distributed [28].

Tack during printing, that is on-press, will vary with press speed, roller speed, ambient temperature and humidity. According to [71] the dimension of tack should be $J\ m^{-2}$ which is equivalent to energy per area. Tack values are mostly given as dimensionless, as in table 3.12, taken from [28].

3.5 The influence of the substrate, printing method and pigment choice on the appearance of the print

Printing substrates range from diffuse to highly specular reflectors. The colour of the substrate also has a huge influence on the appearance of the print, it can change the perceived colour from cold to warm. As an, maybe extreme, example, we would like to discuss the 'amber project' [72]. The aim of this project was to depict the translucency of amber as realistically as possible. We printed with two different sets of inks on white and black paper and used two printing methods, lithography and screen printing, to test the influence of the printing method, the substrate colour and the ink choice on the appearance of the print.

The main difference between lithography and screen printing is the thickness of the ink layer on the substrate. For screen printing it is 15–30 μm [73] and for lithography 1–3 μm [74]. The halftone was in both cases the same, 25 lines per inch (lpi) and rather coarse. It was chosen because we wanted to accommodate three different sets of inks (see [72]) and the lpi was determined by the diameter of the biggest pigment and the mesh size of the screen (55H mesh with a pore size of 105 μm). The print size was 600 mm × 450 mm, to be viewed at a distance of about 4 m. At that distance the halftone is no longer visible. The paper was 240 gms Plike (from G F Smith) in black and white. Plike is made from 100% elemental chlorine free (ECF) wood-free primary pulp, has a very smooth finish and a matt appearance. The CIELAB colour coordinates for white

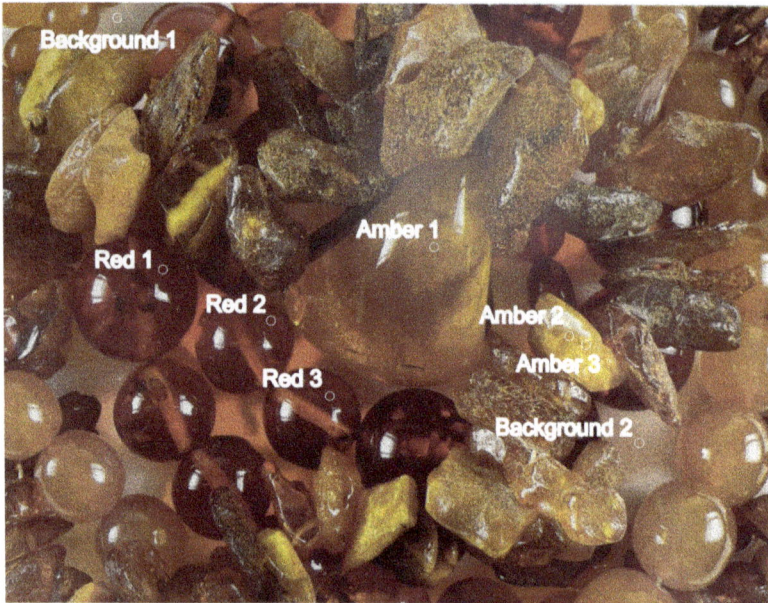

Figure 3.15. Digital image of amber beads, taken with iPhone 12 under studio daylight lighting. The measuring points for the colour coordinates are marked.

Plike are $L^* = 93.99$, $a^* = 2.89$, $b^* = -7.74$ and for black $L^* = 23.25$, $a^* = -.19$, $b^* = -.77$ (CIE Standard Illuminant D50, observation angle 2 deg). The CIELAB, or CIE L* a* b*, colour system represents the quantitative relationship of colours on three axes: the L* value indicates lightness, and a* and b* are chromaticity coordinates. On the colour space diagram (see figure 3.16) L* is represented on a vertical axis with values from 0 (black) to 100 (white). The a* value indicates the red–green component of a colour, where +a* (positive) and –a* (negative) indicate red and green values, respectively. The yellow and blue components are represented on the b* axis as +b* (positive) and –b* (negative) values, respectively. The centre of the colour space is achromatic. The distance from the central axis represents the chroma (C*), or saturation of the colour. The angle on the chromaticity axes represents the hue (h) [75] (figure 3.16).

An image of different amber beads was taken under studio daylight illumination with an iPhone 12 in RAW format (figure 3.15). For classic CMYK (cyan, magenta, yellow and black) printing on white paper the RGB (red, green, blue) colour coordinates of the digital image are translated into CMYK values, halftone and different screen angles, to avoid moiré between the different colour layers, are introduced. The thus-prepared images are then printed on transparent films as positives and either used to expose the prepared screens for screen printing or the metal plates for photolithography.

For lithography we used Van Son Primebio™ process inks made for industrial offset printing (figure 3.18 left). For screen printing, System 3 acrylic paint in cyan, magenta,

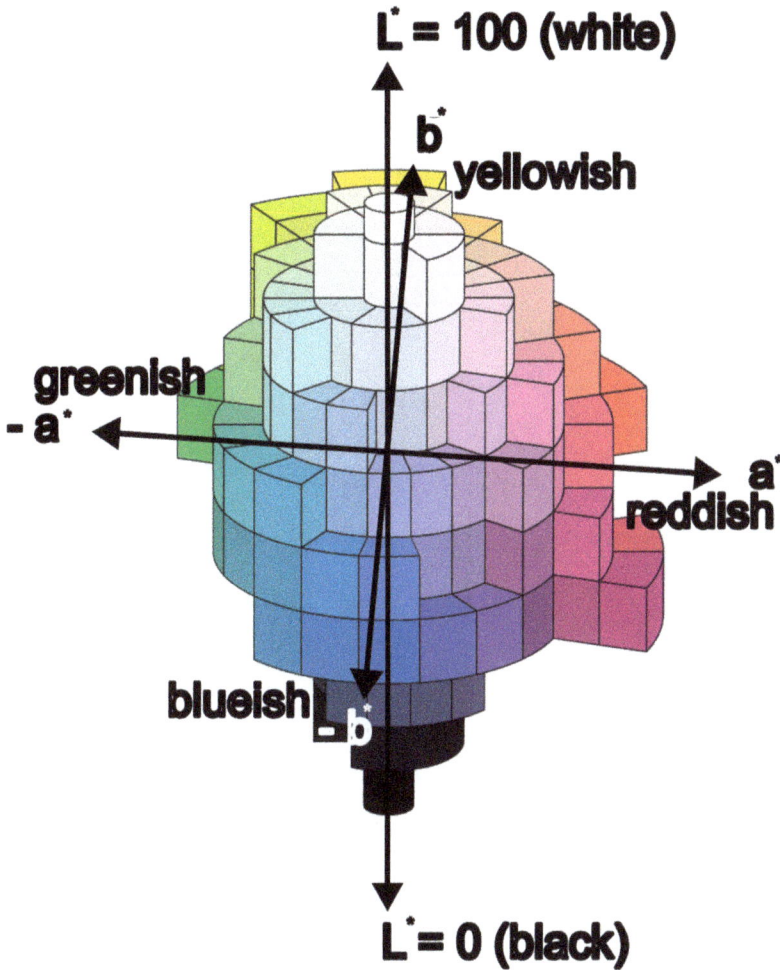

Figure 3.16. The CIELAB colour space. (This image has been obtained and adapted by the author from the Wikimedia website where it was made available by Efa (2022) under a CC BY-SA 4.0 licence. It is included within this chapter on that basis. It is attributed to Efa.)

yellow and black was mixed with Daler Rowney System 3 printing medium (figure 3.18 right). The mixed inks contained 50% paint and 50% medium by weight for CMY and 30% paint and 50% medium for black.

In table 3.13 the CIELAB coordinates are listed for the areas marked in figure 3.15. The measurement themselves were taken from the CMYK screen prints and lithographs on white Plike paper using a colorimeter X-263 Rite i1Profiler. The device uses LED illumination (including UV), has an aperture of 4.5 mm (0.18′) diameter and an illumination spot size of 3 mm 265 (0.12′). The colour measurements were taken in the M1 mode, at an observation angle of 2 degrees. The colour viewing standard ISO 3664:2009 was used for this mode, as the light source contains UV and complies with the spectral distribution specified by CIE illuminant D50.

Table 3.13. CIELAB coordinates for CMYK screen prints and lithographs on Plike white.

	Screen print			Lithograph		
	L^*	a^*	b^*	L^*	a^*	b^*
Amber 3	69.45 ± 4.71	25.67 ± 8.52	59.46 ± 1.72	77.99 ± 1.63	9.22 ± 2.49	52.08 ± 2.54
Red 2	34.73 ± 1.84	32.25 ± 3.61	14.21 ± 0.83	33.48 ± 1.25	41.6 ± 0.39	22.16 ± 1.93
Background 2	63.13 ± 0.37	30.79 ± 0.02	46.60 ± 2.13	72.28 ± 5.82	7.90 ± 0.82	12.61 ± 2.28
Plike white	93.99	2.89	−7.74	93.99	2.89	−7.74

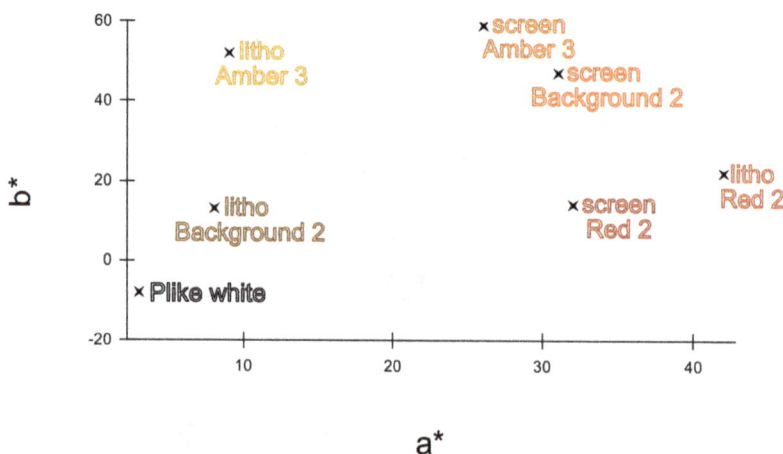

Figure 3.17. CIE a^* and b^* coordinates from table 3.13: CIELAB coordinates for CMYK screen prints and lithographs on Plike white. Positive a^* values signify a redshift, positive b^* values a yellow shift. The negative b^* value for Plike white means that the paper appears bluish white.

The data files include a list of reflectance intensity values in intervals of 10 nm in the range of 380–730 nm.

Table 3.13 and figure 3.17 show that the printing method causes a colour shift even when printed on the same paper. The colour coordinates for Amber 2 and Background 2 show a redshift when screen printed, but Red 2 shows a blueshift. The background colour is much closer to the colour of the paper in the lithograph which is a consequence of a thinner ink layer.

When printing on black paper, we cannot use standard CMYK process inks. Process inks on white are used when the shadows of an image are printed, that is the dark areas in a positive image. Layering of CMY will give black. K (black) enhances the overall contrast but is not necessary when the spectra of the colour filters for recording, either analogue or digital, are perfectly matched to the complementary

Figure 3.18. On the left is the lithograph of figure 3.15, on the right the screen print. Both images are printed overlaying yellow, magenta, cyan and black, 25 lpi, print size 600 mm × 450 mm.

spectra of the printing inks. The highlights in the image have only a very thin ink coverage or none. The white in the image comes from the white of the substrate.

When printing on a black substrate, the shadows are not printed but the highlights are. Since a positive litho plate holds only ink where the plate was not exposed to UV light, that is the shadows in a positive image, for a black substrate the image needs to be a negative. The highlights are then the dark areas on the plate. Printing the highlights means that all inks printed on top of each other should result in white. The print has to be done in additive colours. This is possible when printing with Spectraval™ pearlescent pigments, developed and provided by Merck. Spectraval™ pigments come in red, green, blue and white. They are not pigments in the classic sense since they do not absorb any light but reflect selectively. The particles are mica flakes which are coated in a titanium dioxide (see figure 3.19). Different coating thicknesses change the optical pathlength within the particle and therefore the phase between the light reflected from the top and the bottom of the particle. Different phases influence how the light interacts with itself within the group of all light waves reflected from the particles. The interaction is called interference and the simplest case applicable to the present scenario is thin film interference (see for example, [76] p 286).

Light reaches the eye of the observer when the phase difference between two light waves is

$$2nh\cos(\theta) \pm \frac{\lambda}{0} + m\lambda, \quad m = 0,\ 1,\ 2,\ 3,\ ....$$

Darkness reaches the eye of the observer when the phase difference between two light waves is

$$2nh\cos(\theta) \pm \frac{\lambda}{0} + m\lambda, \quad m = \frac{1}{2},\ \frac{3}{2},\ \frac{5}{2},\ ....$$

The phase difference and therefore the condition for constructive (light) and destructive (darkness) interference depends on:

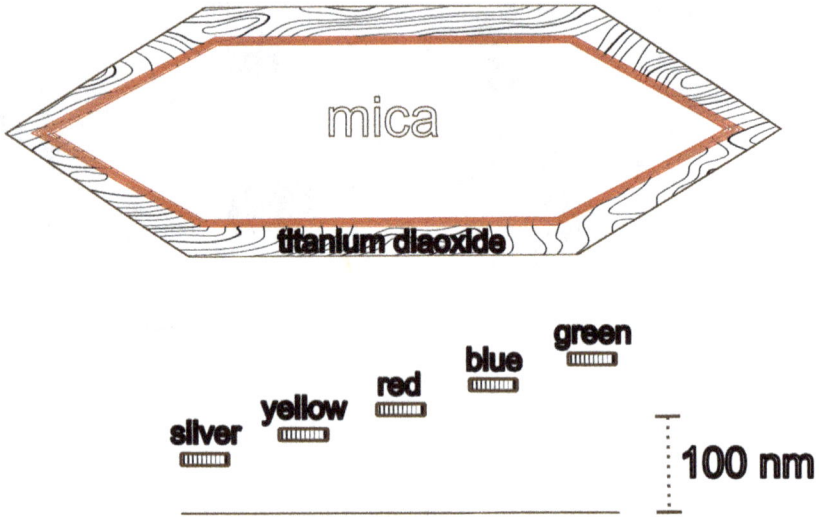

Figure 3.19. Construction of a selectively reflecting pigment. A thin mica plate is coated with titanium dioxide. The thickness of the coating defines the selectively reflected colours in a certain range. The process is very angle dependent and it is important that the particles are deposited flat during printing.

1. The refractive index n of the particle which will be a mixture of the refractive index of mica and the coating.
2. The thickness h of the particle.
3. The viewing angle θ.
4. The wavelength of the light.

Since white light is a mixture of different wavelengths and the refractive index is dependent on the wavelength, different parts of the visible spectrum will interfere constructively or destructively. The thickness of the particle and its orientation in respect to the observer will determine the part of the spectrum which is perceived as colour by the observer. For the strongest colour effect Spectraval™ particles need to be aligned. They need to lie flat on the substrate.

Printing on black paper with selectively reflective pigments creates a very different optical appearance compared to classic CMYK printing (figure 3.20). The colours look washed out but display a change when the print is tilted and create a 3D impression.

Table 3.14 and figure 3.21 show that the screen print is shifted towards red across all measurement points. Because of the smooth surface, the luminance of black Plike paper is quite high and it often appears grey in photographs or when scanned. The colour coordinates of the lithograph are clustered around the white point, $a^* = b^* = 0$, which is not the case when printed on a white substrate with CMYK inks. In comparison with a CMYK print the RGB print looks pale (figures 3.20 and 3.18).

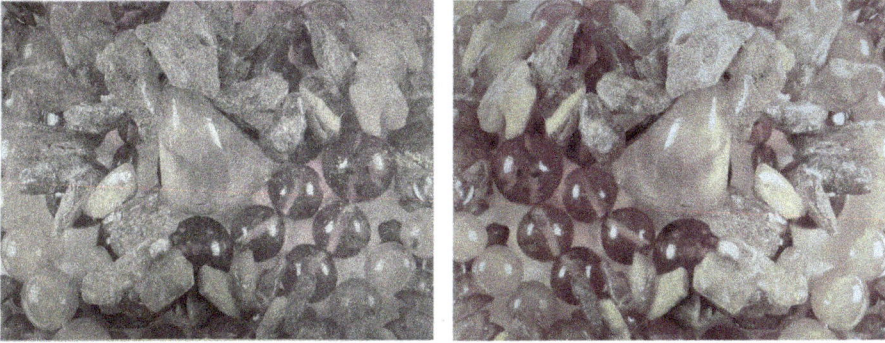

Figure 3.20. On the left is the lithograph of figure 3.15, on the right the screen print. Both images are printed overlaying red, green and blue ink made with Spectraval™ pearlescent pigments on black Plike paper, 25 lpi, print size 600 mm × 450 mm. Because the lithograph is printed as a direct lithograph, the image is flipped.

Table 3.14. CIELAB coordinates for RGB screen prints and lithographs on Plike black.

	Screen print			Lithograph		
	L^*	a^*	b^*	L^*	a^*	b^*
Amber 3	60.37 ± 0.94	3.82 ± 0.12	32.33 ± 0.86	40.45 ± 0.76	0.41 ± 0.31	2.10 ± 0.14
Red 2	46.39 ± 0	8.26 ± 0.23	32.34 ± 0.12	30.90 ± 0	2.43 ± 0	−0.91 ± 0
Background 2	58.21 ± 1.16	4.1 ± 0.27	26.84 ± 1.44	37.94 ± 1.94	0.59 ± 0.35	0.74 ± 0.45
Plike black	23.25	−0.19	−0.77	23.25	−0.19	−0.77

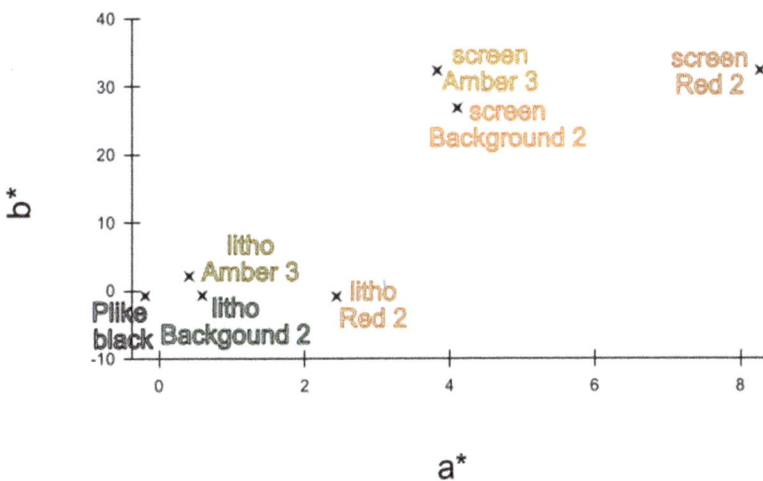

Figure 3.21. CIE a^* and b^* coordinates from table 3.14. Positive a^* values signify a redshift, positive b^* values a yellow shift.

L* on Plike white and L* on Plike black

■ L* on Plike white ■ L* on Plike black

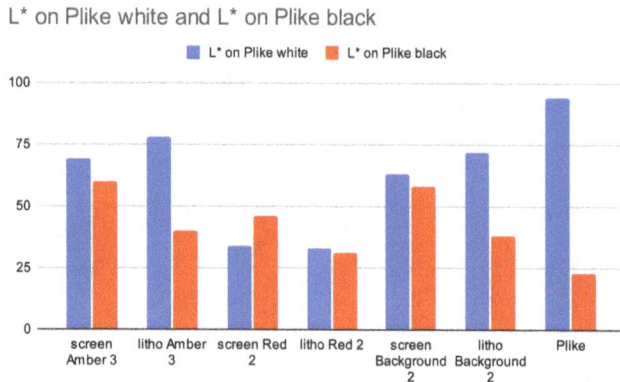

Figure 3.22. Luminance values for the CMYK print on white Plike (cyan) and for the RGB print on black Plike (red).

Figure 3.22 shows that the luminance values for CMYK on white paper are on average higher than the ones for RGB on black paper. This is not surprising. Printing CMYK on white paper means that the luminance of the white background is reduced by the application of colour. The direction is from white to black. Whereas printing RGB on a black background means that the luminance of the black background is increased by the application of colour. The direction is from black to white. The difference in luminance for the amber and the background is greater for the RGB print on black than for CMYK on white. Only for red is it the other way around.

This case study makes it very clear how much the appearance of a print depends on the substrate, printing method, and on the choice of pigments.

References

[1] Rutherford E 1911 The scattering of α and β particles by matter and the structure of the atom *Lond. Edinb. Dubl. Phil. Mag. J. Sci.* **79** 669–88

[2] Bohr N 1913 On the constitution of atoms and molecules *Phil. Mag.* **26** 1–24

[3] Cook N D 2005 *Models of the Atomic Nucleus: Unification Through a Lattice of Nucleons* (Berlin: Springer)

[4] Jones J E and Chapman S 1924 On the determination of molecular fields. II. From the equation of state of a gas *Proc. R. Soc. Lond.* A **106** 463–77

[5] Kendal K 2001 *Molecular Adhesion and its Applictions, The Sticky Universe* (New York: Kluwer Academic/Plenum)

[6] Parsegian V A 2006 *Van der Waals Forces: A Handbook for Biologists, Chemists, Engineers, and Physicists* (Cambridge: Cambridge University Press)

[7] Laurie A P 1937 The refractive index of a solid film of linseed oil rise in refractive index with age *Proc. R. Soc. Lond.* A **159** 10

[8] Medalia A I and Richards L W 1972 Tinting strength of carbon black *J. Colloid Interface Sci.* **40** 233–52

[9] Rubi 2016 *Product Safety Sheet* ECO 88S Ipagsa Industrial S L, Barcelona

[10] Singru R N, Zade A B and Gurnule W B 2008 Synthesis, characterization, and thermal degradation studies of copolymer resin derived from p-cresol, melamine, and formaldehyde *J. Appl. Polym. Sci.* **109** 859–68

[11] Idris O H M, Williams P A and Phillips G O 1998 Characterisation of gum from *Acacia senegal* trees of different age and location using multidetection gel permeation chromatography *Food Hydrocoll.* **12** 379–88

[12] Schulman J H 1941 The oil/water interface *Nature* **147** 197–200

[13] Hande V R and Chakrabarty S 2015 Structural order of water molecules around hydrophobic solutes: length-scale dependence and solute–solvent coupling *J. Phys. Chem.* **B 119** 11346–57

[14] Chandler D, Chandler D L and Wu D 1987 *Introduction to Modern Statistical Mechanics* (Oxford: Oxford University Press)

[15] Ball P 2008 Water as an active constituent in cell biology *Chem. Rev.* **108** 74–108

[16] Enderby J E and Neilson G W 1981 The structure of electrolyte solutions *Rep. Prog. Phys.* **44** 593–653

[17] Soper A K 2000 The radial distribution functions of water and ice from 220 to 673 K and at pressures up to 400 MPa *Chem. Phys.* **258** 121–37

[18] Science and Technology Facilities Council 2020 Empirical potential structure refinement *ISIS Neutron and Muon Source* (Accessed: 10 February 2020) https://www.isis.stfc.ac.uk/Pages/Empirical-Potential-Structure-Refinement.aspx

[19] Chandler D 2005 Interfaces and the driving force of hydrophobic assembly *Nature* **437** 640–7

[20] Dixit S, Crain J, Poon W C K and Finney J L 2002 Molecular segregation observed in a concentrated alcohol-water solution *Nature* **416** 829–32

[21] Swope W C and Andersen H C 1984 A molecular dynamics method for calculating the solubility of gases in liquids and the hydrophobic hydration of inert-gas atoms in aqueous solution *J. Phys. Chem.* **88** 6548–56

[22] Vereshchagin A G and Novitskaya G V 1965 The triglyceride composition of linseed oil *J. Am. Oil Chem. Soc.* **42** 970–4

[23] Berne B J, Weeks J D and Zhou R 2009 Dewetting and hydrophobic interaction in physical and biological systems *Annu. Rev. Phys. Chem.* **60** 85–103

[24] Atgie M 2018 Composition and structure of gum Arabic in solution and at oil-water interfaces *Doctorat* Institut National Polytechnique de Toulouse (INP Toulouse), L'université de Toulouse, Toulouse

[25] Wallqvist A and Berne B J 1995 Computer simulation of hydrophobic hydration forces on stacked plates at short range *J. Phys. Chem.* **99** 2893–9

[26] Huang X, Margulis C J and Berne B J 2003 Dewetting-induced collapse of hydrophobic particles *Proc. Natl Acad. Sci.* **100** 11953–8

[27] Choudhury N and Pettitt B M 2007 The dewetting transition and the hydrophobic effect *J. Am. Chem. Soc.* **129** 4847–52

[28] NAPIM 2017 *NPIRI Printing Ink Handbook* 7th edn (Washington, DC: National Association of Printing Ink Manufactures)

[29] Smook G A 2016 17.1.1 Criteria of performance *Handbook for Pulp and Paper Technologists* 4th edn (Peachtree Corners, GA: Technical Association of the Pulp and Paper Industry (TAPPI)) p 272

[30] Fuchs G and Daugherty J 2008 Formulating printing inks to minimize environmental impact *White Paper* NAPIM https://napimtech.org/Technical/TechIndex

[31] Crivello J V and Reichmanis E 2014 Photopolymer materials and processes for advanced technologies *Chem. Mater.* **26** 533–48

[32] Shukla V, Bajpai M, Singh D K and Singh M 2004 Review of basic chemistry of UV-curing technology *Pigment* **33** 272–9

[33] Decker C 2002 Kinetic study and new applications of UV radiation curing *Macromol. Rapid Commun.* **23** 1067–93

[34] Decker C 2001 UV-radiation curing chemistry *Pigment* **30** 278–86

[35] Allen N S 1996 Photoinitiators for UV and visible curing of coatings: mechanisms and properties *J. Photochem. Photobiol.* A **100** 101–7

[36] Roffey C 1997 *Photogeneration of Reactive Species for UV Curing* (Chichester: Wiley) p 866

[37] Dlugogorski J B Z, Kennedy E M and Mackie J C 2012 Low temperature oxidation of linseed oil: a review *Fire Sci. Rev.* **1** 3

[38] Launer H F 1939 *Determination of the pH Value of Papers* (Washington, DC: U S Government Printing Office) pp 553–64

[39] Matija S *et al* 2004 What is the pH of alkaline paper? *e-Preserv. Sci.* **1** 35–47

[40] Roffey C 1997 *Photogeneration of Reactive Species for UV Curing* (Chichester: Wiley)

[41] 2002 *Handbook of Imaging Materials* 2nd edn ed A S Diamond and D S Weiss (Boca Raton, FL: CRC Press)

[42] Gregory P 1996 *Chemistry and Technology of Printing and Imaging Systems* (London: Blackie Academic and Professional)

[43] Choen K-S *et al* 2010 Dyes and use thereof in imaging members and methods *US Patent Specification* US20080187866A1

[44] van de Hulst H C 1981 *Light Scattering by Small Particles* (New York: Dover)

[45] Feller R L, Roy A, FitzHugh E W and Berrie B H 1986 *Artists' Pigments: A Handbook of their History and Characteristics* (Washington, DC: National Gallery of Art) pp 1–4

[46] CAMEO: Conservation and Art Materials Encyclopedia Online (Museum of Fine Arts Boston, MA 2020) http://cameo.mfa.org/wiki/Main_Page (Accessed: 16 April 2020)

[47] Hudson Institute of Mineralogy *Mindat.org* https://www.mindat.org/a/hudsoninstituteofmineralogy

[48] Patnaik P and Knovel 2003 *Handbook of Inorganic Chemicals* (New York: McGraw-Hill)

[49] Janzen J 1979 The refractive index of colloidal carbon *J. Colloid Interface Sci.* **69** 436–47

[50] Polyanskiy M N 2024 Refractiveindex.info database of optical constants *Sci. Data* **11** 94

[51] National Printing Ink Research InstituteFetsko J M 1974 *NPIRI Raw Materials Data Handbook: Physical and Chemical Properties, Fire Hazard and Health Hazard Data* (Bethlehem, PA: National Printing Ink Research Institute, Francis MacDonald Sinclair Memorial Laboratory, Lehigh University) p v

[52] Koleske J V 2012 *Paint and Coating Testing Manual: Fifteenth Edition of the Gardner-Sward Handbook* (West Conshohocken, PA: ASTM International)

[53] Polymer properties database https://www.campusplastics.com/

[54] Hu L-F, Li Y, Liu B, Zhang Y-Y and Zhang X-H 2017 Alternating and regioregular copolymers with high refractive index from COS and biomass-derived epoxides *RSC Adv.* **7** 49490–7

[55] ChemBK CAS database https://chembk.com/en

[56] Born M A X and Wolf E 1980 Electromagnetic potentials and polarization *Principles of Optics* 6th edn ed M A X Born and E Wolf (Oxford: Pergamon) ch 2 pp 71–108

[57] Born M A X and Wolf E 1980 Chapter IV—Geometrical theory of optical imaging *Principles of Optics* 6th edn ed M A X Born and E Wolf (Oxford: Pergamon) ch 4 pp 133–202

[58] Hoskins S 2004 *Inks* (London: A and C Black)

[59] Kipphan H 2014 *Handbook of Print Media* (Berlin: Springer)

[60] Ek M, Gellerstedt G and Henriksson G 2009 *Pulp and Paper Chemistry and Technology* (Berlin: de Gruyter)

[61] Gross A 1970 *Etching, Engraving, and Intaglio Printing* (London: Oxford University Press)

[62] Cong L 2017 UV curable inkjet ink compositions *Patent Appl.* World WO 2017/031224 Al

[63] Morris D P and Edwards M R 2008 Chemically produced toner and process therefore *Patent Appl.* USA US 2005/0175921 A1

[64] Akiyama H H 2008 Electrostatic latent image developing magenta toner, electrostatic latent image developer, toner manufacturing method, and image forming method. *US Patent Specification* US7422834B2

[65] Marko H L 2012 Method for preparation of a liquid electrostatographic toner and liquid electrostatographic toner *Patent Appl.* Europe 12159498

[66] Ng H T, Ganapathiappan S and Teishev A 2016 Deinkable liquid toner *Patent Appl.* USA US2014/0147172 A1

[67] Kissa E 1999 *Dispersions: Characterization, Testing, and Measurement* (London: Taylor and Francis)

[68] Bingham E C 1922 *Fluidity and Plasticity* (New York: McGraw-Hill)

[69] Öhlund T 2014 Metal films for printed electronics: ink–substrate interactions and sintering *PhD Thesis* Mid Sweden University Sundsvall, Sweden

[70] Green H 1941 The tackmeter, an instrument for analyzing and measuring tack: application to printing inks *Ind. Eng. Chem. Anal. Ed.* **13** 632–9

[71] Hamerliński J and Pyryev Y 2013 Modelling of ink tack property in offset printing *Acta Poligraph.* **2013** 31–40

[72] Trujillo Vazquez A, Fuller H, Klein S and Parraman C 2021 The Amber Project: a survey of methods and inks for the reproduction of the color of translucent objects *Appl. Sci.* **12** 793

[73] Horwood R 1974 Towards a better understanding of screen print thickness control *Electrocomponent Sci. Technol.* **1** 129–36

[74] Flexographic Technological Association 2013 *Flexography: Principles and Practices 6.0* (Bohemia, NY: Flexographic Technological Association)

[75] CIE 2018 Colorimetry 4th edn (Commission Internationale de l'Eclairage) https://cie.co.at/publications/colorimetry-4th-edition

[76] Born M A X and Wolf E 1980 Interference and diffraction with partially coherent light *Principles of Optics* 6th edn ed M A X Born and E Wolf (Oxford: Pergamon) ch 10 pp 491–555